The New
Invisible College

The New Invisible College

Science for Development

CAROLINE S. WAGNER

BROOKINGS INSTITUTION PRESS
Washington, D.C.

Copyright © 2008
THE BROOKINGS INSTITUTION
1775 Massachusetts Avenue, N.W., Washington, D.C. 20036
www.brookings.edu

Library of Congress Cataloging-in-Publication data

Wagner, Caroline S.
  The new invisible college : science for development / Caroline S. Wagner.
     p.   cm.
  Includes bibliographical references and index.
  Summary: "Combines quantitative data and extensive interviews to map emerging global science networks and trace the dynamics driving their growth. Argues that the shift from big science to global networks creates unprecedented opportunities for developing countries to tap science's potential. Offers a guidebook and playbook for policymakers confronting science's transformation"—Provided by publisher.
  ISBN 978-0-8157-9213-0 (pbk. : alk. paper)
  1. Science and state. 2. Science—Study and teaching—International cooperation. 3. Global method of teaching. 4. Research—International cooperation. 5. Information networks. I. Title.
  Q125.W258 2008
  338.9'26—dc22   2008018700

9 8 7 6 5 4 3 2 1

The paper used in this publication meets minimum requirements of the American National Standard for Information Sciences—Permanence of Paper for Printed Library Materials: ANSI Z39.48-1992.

Typeset in Adobe Garamond

Composition by Cynthia Stock
Silver Spring, Maryland

Printed by R. R. Donnelley
Harrisonburg, Virginia

Contents

PART III
TAPPING NETWORKS TO EXTEND THE BENEFITS
OF SCIENCE AND TECHNOLOGY

Foreword

The death of the nation-state has been greatly exaggerated by a series of false prophets over the past decade or so, particularly during the heyday of the information technology (IT) boom of the late 1990s when globalization was picking up steam. Capital, labor, and especially information were said to be unprecedentedly mobile, empowered by new technologies like the Internet; efforts by nation-states to control what flowed across their borders were seen as incredibly retrograde and doomed to failure. In place of traditional governance through the vertical stovepipes of democratic political systems, pundits predicted a flat world of horizontally organized self-governing entities that did away with the need for coercion and the kind of top-down control that we associate with national governments.

The nation-state, it turns out, was more durable than that. Though it was true that technology greatly facilitated the flow of various tangibles and intangibles across international borders, nation-states were able to furnish certain unique services. They alone could wield the "monopoly of legitimate force" that was capable of enforcing rules on a given territory, and they alone had the resources to provide what economists call public goods, goods that private markets will not produce because their use cannot be restricted to a portion of the public. The world, moreover, was not "flat"—the country in whose jurisdiction you lived correlated strongly with a host of relevant life outcomes, like income, health, opportunities for education and employment, and other important measures. In addition, there were public bads as well as public goods, menaces like global disease, terrorism, and political struggles over resources whose effects went well beyond the borders of the countries immediately involved. To cope with them, nation-states continued to be necessary. And finally, not everything could flow across borders seamlessly: countries could continue to impose tariffs, subsidize national champions, restrict the flow of immigrants, and even control access to information and ideas, though the latter was decidedly more difficult. The theory was that the forces of globalization would punish those who violated its rules, so that the system became self-enforcing, but it did not take account of the countervailing political trends that globalization itself produced.

As Caroline Wagner shows in *The New Invisible College*, however, international flows truly challenge the nation-state in certain domains, one of which is the development of modern natural science. Unlike technology, a great deal of research in basic science has the character of a public good: it is hard to exclude people from its benefits, and most important, it can develop only in an atmosphere of free and open exchange. *The New Invisible College* demonstrates in a dramatic fashion the degree to which scientific collaboration has become internationalized. Even though rich countries continue to be the largest sources of funding for scientific research, the character of that research today can be understood only as the by-product of a horizontal process of social collaboration in which merit and results trump any consideration of national origin or jurisdiction.

It is for this reason that the theory of complex adaptive systems and network analysis becomes critical to the understanding of the evolution of science. Modern science is an intensely social process, and like other social systems, it is not planned in a hierarchical manner by national governments. As Wagner clearly shows, it is the emergent characteristic of invisible colleges of researchers who are attracted to one another based on the complementarity of their work. Self-organized systems generate complexity in an unplanned way through the interactions of individual agents; the final result—as in the case of an ecosystem—is much larger than the sum of its parts. Distributions will not be normal; they will be scale-free and follow power-law rules. It is extremely difficult to anticipate in advance where the nodes of scientific discovery will arise or how different researchers will connect with one another. *Weak links, small worlds,* and *nodes* are the most useful terms for understanding the way that scientific discovery advances.

Working from these premises, Wagner points out the following central public policy problem. The development of modern science is without doubt an emergent social process that is international in scope and cannot be effectively controlled by governments. And yet it is the taxpayers of different nation-states who are asked to fund this process. Governments continue to think about scientific research and advancements in national terms, as French or Japanese or American science designed to benefit each particular country. Indeed, much of the impetus for funding scientific endeavors came directly out of the perceived need to promote science as an input into national defense.

But even though the connection between scientific development and national well-being still clearly exists, science itself flourishes best in a world without national borders, where knowledge is not proprietary but flows to those who can push its limits the furthest. And although access to the networked world of modern science is undoubtedly beneficial to poor and developing countries, they are often unable to take advantage of existing scientific resources because of lack of funding and human capital. How can these contradictory imperatives be reconciled?

The first step, this book suggests, is to recognize the nature of the phenomenon of science itself, its networked character, and the degree to which it has been internationalized. Governments and taxpayers in rich countries must understand that this is a domain of extensive externalities, where benefits do accrue, but often not directly as a result of planned investments. Realistically, distinctions must be drawn between different areas of science. Some, like high-energy physics, require large, immobile, capital-intensive investments that necessitate overt cooperation to avoid duplication of effort and to pool funding. Others, like agricultural research, are dependent on locality and need to be dispersed. In particular, developing countries need to avoid simply duplicating the developed world's twentieth-century national science establishments. As they continue to develop, these countries have many alternatives for finding specialized niches that take advantage of the openness of the current system.

Clearly, governance mechanisms will have to evolve to keep pace with reality. The old model of national regulation—in which governments established hierarchical regulatory institutions and international cooperation was undertaken through formal treaty organizations based on national regulators—is of necessity giving way to more flexible forms of governance and cross-border cooperation. Some of this flexibility involves informal cooperation at middle levels of government organization, what Anne-Marie Slaughter labels "intergovernmentalism."[1] Some occurs with the help and participation of nongovernmental organizations (NGOs) and stakeholders directly in the regulatory process. Some involves public-private partnerships among businesses, NGOs, and governments at various levels. All these new forms of governance and international cooperation may seem troubling because, for the sake of effectiveness and decisionmaking speed, they sidestep formal democratic mechanisms of accountability in favor of less accountable and sometimes less transparent mechanisms. But they also seem to be necessary if governance is to keep pace with the speed of the evolution of the social processes being regulated. These are the challenges we as a world community will need to face in the twenty-first century.

As it continues into the future, the internationalization of science will continue to pose other kinds of challenges to cooperation. Scientific research produces not just public goods, but public bads as well: nuclear and other weapons of mass destruction; dangerous biological agents that in the future may be engineered to become more virulent at relatively low cost; and various forms of environmental damage as side effects of good uses. These necessitate control regimes like the one established by the Nuclear Non-Proliferation Treaty. In such instances, though, the same international, networked character of science that makes it work for positive ends also makes it difficult to guard against misuse.

We cannot, however, even begin to deal with either the positive or negative impacts of scientific advancements unless we first understand the nature of the phenomenon we are analyzing. To this end, *The New Invisible College* provides an invaluable service, not only by helping to advance that understanding, but also by shifting the terms of the scientific policy discussion toward a new paradigm necessitated by the nature of the world in the twenty-first century.

FRANCIS FUKUYAMA
Bernard L. Schwartz Professor of International Political Economy
Director of the International Development Program
Paul H. Nitze School of Advanced International Studies
Johns Hopkins University

ACKNOWLEDGMENTS

I spent more than six years conducting research and interviewing individuals to complete this book. The Rockefeller Foundation's Global Inclusion Program (New York) supported the writing, and I am very grateful to the program's director, Janet Maughn, for believing in this project and offering her support and encouragement. Without her, the book would not have been possible.

I wrote much of the book at the George Washington University (GWU) Center for International Science and Technology Policy in Washington, D.C. I am grateful to Nicholas Vonortas at GWU for his support and encouragement and for giving me a home where I could write.

From September 2001 through the spring of 2004, I conducted a great deal of my research as a research project at the Amsterdam School of Communications Research (ASCOR; University of Amsterdam [UvA]) and at the Netherlands Graduate School for Science, Technology, and Modern Culture at Maastricht University. My promoter and collaborator at UvA, Loet Leydesdorff, not only inspired me, but also served as a wonderful guide through the materials, tools, and ideas needed to complete this work. Although many of the ideas in this book can be traced intellectually back to many others, it was Loet Leydesdorff who brought them to life for me. I also thank ASCOR Dean Sandra Zwier for her support.

My editor at the Brookings Institution Press, Mary Kwak, took a personal interest in this book and improved it greatly throughout the preparation process.

Thanks must go to many others for encouragement, support, feedback, and information. Particular thanks to Paul Dufour of the International Development Research Centre (IDRC) in Toronto, Canada. He used his amazing networking skills to connect me to many of the people who gave input to this project, including Jean Woo, IDRC; Geoff Oldham, University of Sussex, Brighton, United Kingdom; Francisco Sagasti, FORO Nacional, Lima, Peru; Calestous Juma, The Belfer Center for Science & Technology, John F. Kennedy School of Government, Harvard University; and Keith Bezanson, International Institute for Sustainable Development, Winnipeg, Manitoba, Canada. While she was a guest editor at the journal *Science & Public Policy,* Josephine Stein (Department of Innovation Studies at the University of East London, United Kingdom) requested the article that began much of this work.

I extend special thanks to the visionary Peter Johnston and his staff in the Evaluation and Monitoring Unit at the European Commission (Belgium). As part of a project for this group, much of the theoretical basis for the argument in the book took shape.

At GWU, colleagues David Grier, Robert Rycroft, Henry Hertzfeld, John Logsdon, Lauren Hall, and Christi Fanelli were very helpful.

At the University of Leuven (Belgium) Policy Research Center for R&D Indicators, Wolfgang Glänzel, Martin Meyer, and their colleagues assisted me in many ways. Thanks also to U.S.-based author Neal Stephenson, who first directed me to the stories from the Royal Society of London and the role of J. Comenius in the formation of the invisible college.

Sylvan Katz, a fellow at the University of Sussex Science and Technology Policy Research Unit (Brighton, U.K.) and an adjunct professor in Mathematics & Statistics, University of Saskatchewan, Saskatoon, Canada, guided me very patiently through the literature on complexity and theory.

During the course of my writing, I had the great good fortune of sharing ideas at a workshop organized by Katy Börner (Indiana University, Bloomington), a visionary leader in the field of visualizing scientific information. She pulled together a number of people to attend a workshop about advancing the field, which was held at the New York Academy of Science on April 4, 2006. The people I met at the workshop helped me to visualize and organize materials for this book. Particular thanks to John Burgoon at Indiana

University, Bloomington, and Bradford Paley of W. B. Paley Illustrations, New York, for the amazing visual of the geographic concentrations of science. Thanks also to Barends Mons from Erasmus University (Rotterdam, Netherlands) for inspiration, and to Dick Klavans and Kevin Boyack, both at SciTech Strategies, New Brunswick, New Jersey, for helping to clarify ways of measuring and visualizing science.

Other friends and colleagues supplied valuable input and insightful comments, including Francis Fukuyama at Johns Hopkins University, Paul Nitze School for Advanced International Studies, Washington, D.C.; Joel Garreau of the *Washington Post;* Pete Suttmeier of the University of Oregon (Eugene), and Philip Shapira and Susan Cozzens from Georgia Tech (Atlanta). Colleagues at the Centre of Research on Innovation and Internationalization (CESPRI-Bologna, Italy) including Stefano Breschi, Franco Malerba, and Lorenzo Cassi assisted with a number of aspects of network theory, as did Jonathan Cave at the University of Warwick (Coventry, United Kingdom). Lucio Biggiero, LUISS University, Rome, offered important comments, and Francisco Sercovich of the United Nations Industrial Development Organization (UNIDO), Vienna, supported and commented on part of the research for several chapters. Thanks also go to Jerry Sollinger of the RAND Corporation (Santa Monica, Calif.) for sharing his unique ability to organize ideas into a good story.

For help on measuring and understanding scientific capacity, I am indebted to Edwin Horlings, Rathenau Instituut, The Hague, Netherlands, and Arindam Dutta and Brian Jackson, my colleagues at RAND.

My family has offered outstanding support throughout the project. In particular, my brother, John Dean Wagner, was unfailingly supportive and enthusiastic. My husband, Dennis McIntosh, along with my children, Julia, Greg, and Nora, offered me unconditional support. I dedicate this work to the memory of my precious sister, Mary Pat Wagner, who died while I was writing this book.

I also appreciate the help of staffers at the Library of Congress and the Royal Society of London for access to rare books.

The New
Invisible College

CHAPTER ONE

The New Invisible College Emerges

> If I am right about the flattening of the world, it will be remembered as one of those fundamental changes—like the rise of the nation-state or the Industrial Revolution—each of which, in its day . . . produced changes in the role of individuals, the role and form of governments, the way we innovated, the way we conducted business . . . the way science and research were conducted.
>
> THOMAS L. FRIEDMAN, *The World Is Flat*

Science—defined broadly as systematic knowledge about the natural world—offers humanity the promise of a better life. Scientific advances throughout history have helped save millions of people from disease, famine, and poverty. The discovery of penicillin, the development of high-yield seeds, and the distribution of electricity are but three examples of the ways in which science contributed to social welfare in the twentieth century. In many countries, such advances have had even more far-reaching effects by spurring economic growth and bolstering the creation of the large and vibrant middle class that many theorists believe is an essential precondition of democracy. Yet other countries have failed to reap similar benefits. Since the birth of modern science in the seventeenth century, gains in and application of scientific knowledge have been unevenly distributed, contributing to the widening gulf between the developed and developing worlds.[1] In this book, I seek to explain why that is so. I also examine how science is changing and how a new framework for the governance of science can help bridge the gap between the scientific haves and have-nots.

Thomas L. Friedman, *The World Is Flat: A Brief History of the Twenty-First Century* (New York: Farrar, Straus and Giroux, 2005).

As Thomas Friedman notes in the epigraph to this chapter, the organization of science is changing in fundamental ways. These changes are both less and more extensive than he suggests. Despite the accelerating diffusion of scientific data, information, and knowledge, the world of science remains far from flat. But its focus has changed from the national to the global level. Self-organizing networks that span the globe are the most notable feature of science today. These networks constitute an invisible college of researchers who collaborate not because they are told to but because they want to, who work together not because they share a laboratory or even a discipline but because they can offer each other complementary insight, knowledge, or skills.

These networks link scientists working in faraway countries through virtual ties. They also organize the constant physical churn of researchers around the world. They furnish the framework within which research teams form, mutate, dissolve, and reform, bringing together scientists from diverse backgrounds and sending them out again with new knowledge to share. In the twenty-first century melting pot of science, national citizenship or allegiance plays a minor role. Scientific curiosity and ambition are the principal forces at work in the new invisible college.

In contrast, scientific nationalism—in which countries view scientific knowledge as a national asset—was the dominant model of the twentieth century. National science ministries and policies funded and controlled scientific research to advance domestic goals, such as economic prosperity and military strength, and opportunities for collaboration were often constrained by national rivalries. This approach paid large dividends for those countries with the wealth, resources, and culture to invest in, retain, and build on a wide range of advances in knowledge. But it left many countries, which collectively represent the majority of the world's population, out in the cold.

The rise of the new invisible college creates challenges and opportunities to promote social welfare and economic growth. In particular, it gives developing countries a second chance to create strategies for tapping into the accumulated store of scientific knowledge and applying what they learn to local problems. This book seeks to lay the groundwork for such strategies by describing the new invisible college and explaining how it works. By applying insights gained from recent advances in network theory, I present a framework for understanding the organization of twenty-first century science.[2] I use a mix of quantitative and qualitative data to describe global networks and identify the rules that fuel their operation and growth. This information serves as a basis for discussing the policy challenges posed by the rise of networked science. Science policy can no longer be made based on national borders, even though nations still play an important role in promoting and regulating scientific activity. Because the structures that create knowledge are not contained within nations, they cannot be managed within national borders. New principles must guide policy now and into the future.

Origins of the New Invisible College

Even though today's invisible college is very much a twenty-first century phenomenon, it also represents the reemergence of an old idea. A review of history shows that the invisible college is not new to science—the same term was used to describe the group of like-minded independent scholars who first pioneered observation and experimentation to study nature in the seventeenth century. Science in those early days was the work of natural philosophers, usually those of independent means like Sir Isaac Newton and Irish chemist Robert Boyle. These individuals, who were largely free from government influence, shared information and insight in a universal language (Latin) without regard for disciplinary boundaries (which at the time barely existed). Then as now, networks characterized scientific organization and inquiry, with the early scientists corresponding and exchanging ideas as part of a common search for knowledge.

As the centuries passed, science progressed a long way from its roots. It became increasingly professionalized. Laboratories, such as those led by Pierre and Marie Curie, formed to focus on specialized subjects like biology, astronomy, physics, or medicine. A process of nationalization followed as nation-states consolidated in the nineteenth and twentieth centuries. Governments began to expand their authority over scientific activity by creating national scientific establishments, such as France's Centre National de la Recherche Scientifique (CNRS). Established in 1939, CNRS now manages more than 1,000 research groups across the country.

Strategic rivalries and fierce economic competition spurred similar and redundant investments in "big science" in other countries, particularly in the wake of the two world wars, which showcased the ability of science to bolster military strength. National governments grabbed hold of science and made vast investments in both military and civilian research establishments. Institutions like the U.S. National Science Foundation and the Russian Academy of Sciences invested heavily in the basic sciences. Sister agencies invested in high-profile projects—such as the race to the moon and the fight to find a cure for cancer—designed to cultivate an aura of national strength and prestige.

More recently, the structure of science has been changing yet again, with the rise of the new invisible college. Five forces that are driving the shift in the structure of science can help us understand this important development:

—**Networks.** Networks are made up of connections among scientists. The connections can exist within formal institutions or established projects, but they do not stop there. They are forged through meetings and common interests and extend across vast geographic distances. These networks are not designed or dictated by anyone, but neither are they random, and they operate according to underlying rules and dynamics that differ from those that have governed the organization of science since at least World War II. Armed with an understanding of these dynamics, policymakers will be better positioned to take advantage of the invisible college's strengths.

—**Emergence.** Networks among scientists emerge in response to new information, new connections, and new opportunities. Science is not a command-and-control system; it has more in common with an ecosystem than with a corporation. New ideas emerge from the combination and recombination of people and knowledge. Researchers with the freedom to identify the people and tools that can advance their work organize themselves into groups. Emergence, as a powerful force in knowledge creation, should be harnessed and nurtured in our networked era.

—**Circulation.** Brains circulate. Trained researchers move to places where they can maximize their access to resources and best contribute their talents to the pool of scientific knowledge. Knowledge and information also circulate. Unexpected connections arise from data placed on the Internet or otherwise shared among researchers. Researchers often do not even know that they will find a data set useful until they stumble across it in the well-known serendipity of scientific discovery. By promoting the circulation of people, information, and ideas beyond political borders, the invisible college can advance knowledge accumulation more effectively and efficiently than scientific nationalism.

—**"Stickiness."** Place (location) still matters in science and innovation.[3] Despite the influence of the information revolution, face-to-face meetings remain essential. Beyond this, some sciences require large-scale, expensive equipment to advance research, making them sticky in comparison with other fields. Others require resources that are available only in certain locations. This stickiness promotes the geographic clustering or concentration of scientific activity. These clusters can become extremely productive because of the convergence of resources, people, and ideas. Clustering is an essential feature of the knowledge system, and even though scientific research is being distributed across the landscape, policy must also make room for specialization.

—**Distribution.** Once the realm of the lone genius, science is now a contact sport. Scientists and engineers around the globe increasingly see the benefits of coming together in teams that rely on distributed tasking.[4] Such collaboration is made possible by the vast growth in scientific capacity over the past century, as well as the growth of Internet-based technologies. The increased reliance on distributed tasking means that researchers no longer have to be in the same place as their collaborators, nor do they have to be in the same place as the problems they seek to solve. This trend toward distribution creates new opportunities for scientists and policymakers to access knowledge wherever it can be found.

A Burst of Discovery

Some of these forces and their effects are exemplified by a project called BeppoSAX.[5] Initially designed as a collaboration between Italian and Dutch astronomers to study the birth of the universe, this effort eventually brought together astronomers from around the world.

Gamma rays—bursts of light released when a star dies—are one of the most important sources of data on the nature of the universe's early moments. Before the inauguration of the BeppoSAX project, gamma rays were seen only rarely, on the chance occasions when they were caught by satellites whose primary mission was to scan the earth for unannounced nuclear bomb tests. The Explorer 11 satellite, launched in 1961, was the first to carry a gamma ray telescope. It picked up 22 gamma ray events in its four months of operation, offering a tantalizing glimpse of the additional data that might be collected by a satellite dedicated to this goal.[6]

In the early 1990s, a group of Italian astronomers at the Istituto di Astrofisica Spaziale e Fisica Cosmica di Roma launched a project to build a satellite devoted to scientific x-ray observation, independent of the spy mission of the Explorer satellites. The project quickly attracted international support with the participation of researchers at the Stichting Ruimte-Onderzoek Nederland (the Netherlands Institute for Space Research). As Marco Feroci of the Istituto di Astrofisica Spaziale observed:

> Making a satellite is very expensive, so you have to do the best job you can right from the start. To get the funds, you must convince the political powers that you are doing the best science. We had a team of Italian astronomers from all over Italy, but we needed expertise in the instrumentation. The Dutch were the best people doing this kind of instrumentation at the time we started our project, so we linked up with them.[7]

Several Italian and Dutch firms collaborated on the construction of the BeppoSAX satellite, which was launched in April 1996 and operated until April 2002.[8] In addition, the BeppoSAX team created a worldwide network of researchers to follow up on the data collected by the satellite. As Luigi Piro, the director of the BeppoSAX project science office, explained: "This experiment was designed from the beginning to be a network. Fifty observatories participated. This way, we could share data very quickly. Exactly as results came in as to the location [of a gamma ray event] we would send information to the e-mail network. Any person who had results or saw something could e-mail the others."[9] In this way, BeppoSAX operated as both a social network of researchers and a technical network using the Internet to link scientific equipment.

As the BeppoSAX team accumulated data on gamma ray events, its members became increasingly attractive collaborators to astronomers working on related questions. The researchers were soon overwhelmed with requests to coauthor papers. In deciding which requests to accept, Piro explained, the team relied on two factors:

> We looked at both quality of their data and their reputation. These are both basically the same. We went first to the highest reputation in the field because this basically guaranteed the quality of the data. We had data on gamma rays and other groups had complementary data on other wavelengths. We could not check all their data! So we depended on their good reputation to ensure that they had good data. We shared our information and they shared theirs and we gained a broader understanding of our results.[10]

He added, "As we worked with people from all over the world, we developed trust. It developed over time. This is what led to successful collaborations." These collaborations led to the publication of some 1,500 articles drawing on BeppoSAX data.[11]

This productivity was facilitated by the BeppoSAX team's unusual decision to make its data freely available to anyone. In the past, the Istituto di Astrofisica Spaziale—a government-funded institute—had sought to protect and secure any data it obtained. In the early stages of their project, however, Piro and his colleagues became convinced that open distribution was the best way to promote their own research. As Piro explained:

> We discussed this quite a bit in our team. We wanted to share data, but we weren't sure how to do this. A researcher has the right to explore his

own instruments or research, but the entire community should also benefit. So we decided that we would waive our rights to have the data [on the location of gamma rays] to ourselves and in turn share them right away. This was not the standard practiced at that time. But it was a good decision because it allowed research that led to a harvest of data. As we shared, others shared with us, too.[12]

As this decision shows, the BeppoSAX team was picking up signals not just from gamma rays, but also from the transformation of science beyond the nation-state. Collaboration on providing equipment, reliance on a far-flung network, widespread distribution of data, the need for coordination, the value of openness—all are hallmarks of the new invisible college.

Early Reactions to the New Invisible College

The rise of self-organizing social networks among scientists, such as the BeppoSAX gamma ray team, is remapping science across the globe and changing the rules by which it is conducted. But in advanced countries, policymakers have been slow to grasp the importance of these networks. In the United States at the end of the 1990s, most policymakers were only vaguely aware of the global science system. Perhaps because of the enormous size of the U.S. science system, many saw little need to pay attention to the emerging global science and technology system. Others saw international science as an appendage of U.S. science. In an attempt to explain why U.S. science agencies did not need a global strategy, one congressional staffer commented, "International science is just foreign aid in another form."[13] Recommendations to U.S. agencies that they should take advantage of and nurture international scientific cooperation generated little interest. The U.S. Department of State's lack of attention to international science continued to be a topic of discussion at science policy meetings and among those involved in science advocacy, but the marriage of science and foreign policy remained an elusive partnership.

The European Union (EU) took a different approach. Beginning in the late 1990s, it sought to encourage cooperation in science among member states as part of its efforts to create a European Research Area (ERA).[14] Participants in a series of framework programs for research and technological development identified thematic priorities, such as broadband research or transportation, while making funding conditional on collaboration involving two or more EU countries. As a result, scientific collaboration grew rapidly in Europe. Little attention was devoted, however, to partnerships reaching beyond the EU's borders.

The inward-looking focus of the highly advanced scientific establishments did not go unchallenged. Calls from the United Nations and the World Bank to harness science for development became increasingly common in the 1990s, and programs were put in place to counter such phenomena as "brain drain" and the "digital divide"—terms that implied a win–lose structure to scientific knowledge. At the same time, other organizations focused on building scientific capacity within developing countries and encouraging links to research institutions in the developed world.

None of these policy approaches paid particularly large dividends. U.S. policymakers found themselves under increasing pressure to promote international scientific collaboration, and a great deal of diplomatic effort was expended to negotiate science and technology agreements. These agreements, though, had very little impact on the actual direction of cooperation. EU policymakers found that projects intended to strengthen the ERA often included non-EU members, which caused confusion about where the research benefits were accruing. Development agencies found that efforts to build science and technology capacity did not stick well in poor countries, and attempts to construct links between developed and developing countries presented difficult challenges.[15]

For the most part, efforts by developing country governments to imitate the infrastructure and investments of scientifically advanced nations have also failed, often because of flawed policy design. For example, many governments have created distinct policies for industry, for science, and for education, each designed to generate new knowledge or solve local problems, but with little cross-referencing among them. The policies are often written by three different ministries (usually communications, industry, and science/education) with minimal incentive to coordinate with one another. In addition, as part of their science and technology policy, many governments have established lists of priority areas for investment. Unfortunately, these lists tend to be generic, reflecting "hot" areas in global science—such as biochemistry, genetics, or nanomaterials—rather than an effort to link priority investments to local problems and issues. A similar lack of realism hampers many efforts to create national innovation systems. The plans themselves may be attractive, but they often have very little backing in the form of political will or budgetary allocations.

Even more important, these plans are typically ill conceived. For the most part, they propose to build a national system along twentieth-century lines, instead of reflecting the emerging system of science and technology that will be the environment within which nations must compete for talent, resources, and funds in the twenty-first century. Such efforts are doomed to fail because they neglect to account for the shift from a nationally centered scientific system to a global one in which researchers, not national authorities, set the rules. This shift presents new challenges for governments, who exercise less control over science than they did in the heyday of big science. But it also creates new opportunities, particularly for leaders in developing regions who sense the importance of the rise of the invisible college. Some policymakers are now turning away from creating national innovation systems and moving toward establishing knowledge systems that scan for knowledge globally and tie down knowledge locally. The shift is away from a focus on building institutions and toward a focus on the functions that further knowledge and adaptability.

The renewed influence of the networked model of science is very good news for developing countries in this sense—the global network is an open system that offers opportunities to new entrants, notably countries that did not actively participate in the system in the twentieth century. But the network is not transparent. It may operate by unwritten rules, but it operates by rules nonetheless, as well as by norms that are not controlled by any institution or government agency. No political official can promise membership in the new invisible college, but learning the rules, norms, and mechanisms that govern networks can improve policy outcomes.

Organization of This Book

To govern the emerging invisible college properly and extend its benefits to formerly excluded places and people, scientists and policymakers need to understand its principles. This book, then, focuses on describing and illustrating the five factors that are shaping the landscape of early twenty-first century science. Using theory and example, I make the case for a science policy that treats science and technology as an emergent networked system rather than as a national asset.

Part 1, "Rethinking Science and Technology as a Knowledge Network," reinterprets the organization of science as a set of emerging global networks instead of a set of nationally controlled institutions. Chapter 2 underlines the magnitude of this shift by describing the systems that evolved through history—systems that revolved around nations, not nature—and are now being left behind. Today, global scientific collaboration increasingly aims not to serve the interests of nations but the creation of knowledge. Although collaboration is spreading across all fields, it takes different forms in different disciplines. Four types of collaborative activity, each described in chapter 2, can be identified: coordinated, geotic, megascience, and distributed.

Chapter 3 identifies the factors and forces that drive the emerging structure of twenty-first century science, drawing from recent findings in physics, biology, and social theory. Increasingly science operates as a set of complex adaptive networks at the global level. These collaborative networks do not form randomly. They emerge from the choices of hundreds of individuals seeking to maximize their own welfare, and they exhibit identifiable regularities, much as markets do. Notably, weak ties, small worlds, redundancy, reciprocity, and preferential attachment interact to influence the pathways for the flow of knowledge as well as to shape the growth and evolution of networks. By understanding these forces, we can also learn how best to make use of the associated networks.

Part 2, "The Labyrinth of the World: Understanding Network Dynamics," details the dynamics of the new invisible college. Chapter 4 uses quantitative data to establish that global science does indeed operate like a network and that this network is growing at a spectacular rate. The chapter also explains how the network expands by focusing on the motivations that drive the individuals who constitute it. It shows that the pattern of collaboration in a wide range of disciplines follows the scale-free distribution that is characteristic of complex adaptive systems and explores the simple rules that generate such complexity. It also discusses the role of circulation in the new invisible college and investigates its implications for developing countries by focusing on the difference between "brain drain" and "brain gain."

Chapter 5 turns away from people to discuss the role of place in the invisible college. Even though new technologies make it possible to transcend geographic boundaries in a way that was almost unimaginable to previous generations, the geography of the invisible college is not entirely virtual. Place still matters. Chapter 5 shows why and suggests how that should affect our thought process about distributing scientific resources and devising strategies to more broadly diffuse their benefits. Science is currently highly concentrated in advanced countries, partly because of political support and partly because of the cumulative advantage of place. Such concentration can sometimes be advantageous and even necessary, but in other situations, distributed facilities or partnerships are more appropriate. Chapter 5 also suggests ways to redesign science policy to yield a more equitable distribution of the benefits of scientific knowledge. Finally, the chapter introduces the concept of a dual strategy that calls for both "sinking" of investments and "linking" to the global network.

Chapter 6 builds on this discussion by addressing the issues of scientific capacity and infrastructure, which are prerequisites for participating in the new invisible college. It defines capacity and analyzes the institutions and functions that constitute its essential underpinnings. It also considers alternatives to the model of scientific nationalism, which required each nation to build its own scientific infrastructure. Today, even though the core elements of scientific infrastructure must be available locally, they need not all be provided by the national government. Alternatives include sharing at the regional or international level.

Part 3, "Tapping Networks to Extend the Benefits of Science and Technology," places the findings about the global system into a governance framework. Chapter 7 presents policy recommendations for both advanced and developing countries. The chapter makes the case for a comprehensive science policy aimed at governing the new invisible college in a way that will more broadly disseminate its benefits to those formerly excluded from full participation. New mechanisms for supporting and using science will need to be crafted in response to this shift. And the skill of policymakers in crafting such strategies will largely determine who emerge as winners and losers from this period of complex change.

The new social structure of science poses significant, although divergent, challenges for both advanced and developing countries. Advanced countries will have to redefine their roles so that they no longer see themselves as "donors" but as participants in a global system. Developing countries have a unique opportunity to take advantage of the changing system by linking to the network and then tying knowledge down locally. Policies based on two key principles—open funding and open access—can help developing countries achieve these goals.

The tremendous influence of science on our global social and economic development raises the stakes for understanding its structure and dynamics. More than 50 years ago, science historian Herbert Butterfield predicted that the history of modern science would acquire an importance commensurate with anything that has come before in our study of the human condition. Science, he argued, "is going to be as important to us for the understanding of ourselves as Greco-Roman antiquity was for Europe during a period of over a thousand years."[16] The pace of change is so quick, however, that we cannot wait for historians to work it out. We must understand the system as it is unfolding, and that is the purpose of this book.

A Note about Methodology

I conducted research for this book over several years and employed both qualitative and quantitative methods.. The research proceeded from the broad to the specific—beginning with an analysis of the global network of science, moving to an examination of networks within disciplines of science, and then exploring the methods and motives of communication among individual scientists. Originally, one question drove the research: Why is international collaboration in science growing at such a spectacular rate?

I chose the methods for answering the original question for their ability to reveal the dynamics of interrelationships at the global level. The dynamics within any social network are difficult to measure, and global science is no exception to this rule. Communications reflect the dynamics of the global network, but they take place at different levels of formality and can be ephemeral.[17] As a result, this study focuses on activities of a more formal nature—those revealed in published articles—as opposed to conference participation, which is much less formal and of a shorter duration. For the most part, any scientific communication can be identified only by the traces it leaves behind (such as co-authored papers). In addition, only codified or published communications can be quantified.[18] But even though such data might represent no more than the tip of the iceberg of all scientific communications, they can be rich in information. For example, by identifying the home institutions of authors, we can show how knowledge production is increasingly distributed around the world. And because articles tend to report the results of successful communications, the overall study was biased toward collaborations that produced outcomes of enough substance to gain attention at the level of formal, peer-reviewed publications. All publications data were drawn from the Institute for Scientific Information (ISI) databases.

In addition to analyzing this type of quantitative data, I interviewed dozens of scientists and engineers who actively participate in global collaborations. Because I conducted interviews with those who are highly active in international collaboration, their stories reflect success in long-distance communication and networking. These interviews revealed some of the reasons that researchers choose to collaborate across geographical distances and across disciplines, as well as the challenges they face. The interview process shed light on why researchers study and work outside their home country. The interviews also yielded informative examples of how collaboration can create innovative approaches to research. I draw on these interviews at several points in the book to illustrate and amplify the conclusions generated by the data.

I

RETHINKING SCIENCE AND TECHNOLOGY AS A KNOWLEDGE NETWORK

The great body of science, built like a vast hill over the past three hundred years, is a mobile, unsteady structure, made up of solid enough single bits of information, but with all the bits always moving about, fitting together in different ways, adding new bits to themselves with flourishes of adornment as though consulting a mirror, giving the whole arrangement something like the unpredictability and unreliability of living flesh. Human knowledge doesn't stay put, it evolves by what we call trial and error, or, as is more usually the sequence, error and trial.

LEWIS THOMAS,
"On the Uncertainty of Science,"
Harvard Magazine 83, no. 1 (1980): 19–22.

CHAPTER TWO

THE TOPOLOGY OF SCIENCE IN THE TWENTY-FIRST CENTURY

> People, social actors, companies, policymakers do not have to do anything
> to reach or develop the network society. *We are in the network society,*
> although not everything or everybody is included in the networks.
>
> M. CASTELLS AND G. CARDOSO, *The Network Society*

S cience operates at the global level as a network—an invisible college. In contrast to the operations of science at the national level, where agencies manage and policy directs investment, no global ministry of science connects people at the international level. Yet most scientists collaborate with colleagues in other countries. The more elite the scientist, the more likely it is that he or she will be an active member of the global invisible college. This chapter explores why and how the invisible college began, how it is organized today, and why understanding it is important for governing science in the twenty-first century.

But first let's take a quick look at a case in point.

From Bayreuth to Brazil

In the spring of 1997, Wolfgang Wilcke, one of the world's leading soil scientists, left his research laboratory in Bayreuth, Germany, and boarded a plane for Brazil with a doctoral student, Juliane Lilienfein.[1] Although they would

M. Castells and G. Cardoso, *The Network Society: From Knowledge to Policy* (New York: Center for Transatlantic Relations, 2006).

soon need to speak Portuguese, they continued to chat in their native German as the plane carried them to their destination. The topic at hand was how to coordinate a scientific research project with three different project leaders across eighteen sites and the Atlantic Ocean. Wilcke and Lilienfein were traveling to Brazil to undertake the somewhat improbable task of studying dirt. More to the point, they were setting up a research project to learn how agricultural systems could be managed to reduce nutrient losses and close nutrient cycles. Their ultimate goal was to find out how farming could be better sustained in poor soils.

It was not that Germany did not have plenty of dirt to study. It did. But Brazil had two particular attractions. One was the deeply weathered tropical soil of the Brazilian *Cerrado,* a region of woodlands and savanna covering more than 2 million square meters in central Brazil. The other draw was Lourival Vilela, a local professor who also loved dirt. Vilela was leading a research team working on soil questions similar to Wilcke's at the Brazilian Ministry of Agriculture (commonly called Embrapa), within a state-funded Brazilian agricultural research institute. Wilcke's research director had introduced the two men at a 1996 conference, where they discovered a shared interest in comparing nutrients in the soil under different farming conditions. Their discussions led to the collaboration that brought Wilcke to Brazil.

Sustainable farming conditions had become a hot topic in environmental studies in the 1980s and early 1990s. Wilcke knew that the collaborative research project would be stimulating and could result in publishable findings. He also guessed that the challenge of farming in tropical soils would attract donors. This hunch paid off when Wilcke obtained funding from a German government agency. Vilela also approached his government for funds, and the project attracted additional support from a third source—the World Bank's program focusing on soil, water, and nutrient management. Armed with funding for their research, Wilcke and Lilienfein began to make plans to visit the Cerrado, where farmers grow soybeans in difficult tropical soil.

Wolfgang Wilcke was first attracted to the field of soil science by the complex nature of soil, in which water, oxygen, organisms, and plant life all interact. As a graduate student, after beating out hundreds of other applicants to gain entry to a program in environmental studies at the University of Mainz, Wilcke took classes in atmospheric studies, plant and earth science, and hydrology (the study of water). He also found the chance to apply both chemistry and biology appealing. In addition, studying soil offered the possibility of travel to fascinating places because soil is highly location-specific—soils in

Europe, for example, differ greatly from those in the tropics. In short order, Wilcke's enthusiasm for soil research and talent for obtaining research funds earned him the highest title in the German academy: Professor Doctor.[2]

In science, distinction in research, coupled with the ability to attract funding and collaborators, results in a prized commodity—freedom to pursue one's own curiosity. In Wilcke's case, this freedom led him to explore such questions as the following: Why does farming in tropical regions have such poor outcomes? (Even under several different farming methods, including tilling, no tilling, and use as pastures, farms in these areas have very low output compared to farms in more temperate regions.) Why must people in these regions—where plant life is lush and rain is abundant—scratch a living from a hardscrabble soil? Why do farmers still use ancient methods of farming in places where the fields are flat and the soil is firm enough to tolerate agricultural machinery?

The Brazilian Cerrado was a good place to research these questions. The region receives five times as much rain as the fertile valleys of Germany, making the Cerrado a particularly interesting place to study soil. In Germany, rain is a blessing. But in tropical Brazil, years of heavy rain drain nutrients from the soil, and this weathering process is so intense that, over time, the soil becomes acidified. At this point, the organisms, minerals, moisture, and chemicals that can make soil rich simply disappear. The process is even more intense in the inner tropics where the amount of rain can be double that received by the Cerrado. When the rain forests in these regions are cut down for farmland, the soil, which draws nutrients from the plant canopy, is exposed to more weathering and becomes damaged even faster.

Nutrient-poor soil like that found in the Cerrado can be farmed, of course. But to be productive, local farmers use chemical fertilizers when they can afford them. These fertilizers are difficult to obtain and dangerous to handle, and farmers find that they need more and more fertilizer to get the same yield. The depleted soil simply cannot retain the nutrients in the fertilizer, so they quickly wash into the groundwater. People then drink this contaminated groundwater, which exposes them to risk factors for potentially fatal diseases like cancer.

Wilcke and Lilienfein spent weeks in the Cerrado working with Vilela's team. They studied tilled fields, took soil samples from nontilled farms, and collected water samples and information on farming methods. The trip was fruitful. The team produced important findings in soil science, concluding that conventional methods of soil management and farming had negative effects on the soil. With the new information, the team was able to offer recommendations to help improve local farming. In addition, Wilcke and his Latin American colleagues published several articles in scientific journals.

But Wilcke probably did not immediately consider another result of his Brazilian trip—the strengthening of his membership in the centuries-old fellowship of the invisible college. Although Wilcke and Vilela did not work for the same university or draw funds from the same ministry, they conducted a joint research project in the backfields of the Brazilian Cerrado that spanned two years. In doing so, they created links in a knowledge network. Although the links themselves are invisible, the connection they create is real, and that connection leads to real outcomes: new knowledge and, in this case, tangible improvements in the lives of Brazilian farmers and the people they serve.

Origins of the Invisible College

Such connections are not new. Links among researchers that extend beyond particular institutions and places have been known as the invisible college at least since 1645, when the Irish scientist Robert Boyle (often called the "father of chemistry") used the term in a letter to his tutor. Boyle was describing the interactions of a small group of like-minded natural philosophers, also known as the "virtuosi." Boyle wrote in a time of intellectual and social ferment. By the mid-seventeenth century, the improved telescopes of Galileo and other early astronomers had yielded more precise measurements of the movement of heavenly bodies. These movements were shown to follow predictable patterns that could be uncovered through scientific observation and study. These discoveries challenged the Aristotelian tenets that the heavens were not only divine and immutable, but beyond human understanding.

As the Aristotelian worldview began to give way, interest in the empirical exploration of nature spread across Europe. In the mid-seventeenth century, scientific societies and academies were established almost simultaneously in five European cities.[3] These societies were intended to facilitate the communication of ideas, the formulation of experiments, and the sharing of results, increasingly through the printed word. Between 1630 and 1830, at least 300 scientific journals were launched.[4] The growth rate of scientific literature was exponential, with the number of scientific journals growing by a factor of ten about every fifty years since.[5]

Boyle's invisible college included such notables as biologist Robert Hooke; mathematician William, Viscount Brouncker; the Reverend John Wilkins, a future head of colleges at both Oxford and Cambridge; and Sir Christopher Wren, the accomplished astronomer and architect of St. Paul's Cathedral.

This invisible college emerged at a time of great political strife in England. The civil wars, which began in 1642 and raged for much of the following decade, split Britain into two camps: the parliamentarians, who sought to defend Parliament's traditional role in matters such as taxation, and the royalists, who favored a stronger monarchy. But the early experimentalists, who as individuals held divergent political views, set aside these differences to pursue their shared interest in studying the "sensible realm" through experimentation. Eventually their discussions gave rise to the Royal Society of London, the oldest scientific society in continuous existence.[6]

During a time of political revolution and civil war, the future members of the Royal Society were revolutionaries of a different kind. They asked basic questions about nature that challenged the religious and academic orthodoxy of the time. Following private introductions, they initially met informally and corresponded occasionally. In the late 1650s, a group within the invisible college began to meet more regularly at Gresham College in London. In 1660, after an inspiring lecture by the polymath Sir Christopher Wren, the gathered group decided to form a "College for the Promoting of Physico-Mathematical Experimental Learning."[7] They took the motto *Nullius in Verba* ("on the word of no one") to show that they were prepared to test and seek verification of facts rather than accept received wisdom.[8]

Today, when questioning received wisdom is a cultural norm, it is difficult to fully grasp the audacity of these experimentalists in publicly declaring their allegiance to science. But their gamble paid off. In early 1660, after a republican interregnum lasting eleven years, the British monarchy was restored. The new king, Charles II, took an interest in the group's work, largely because of his friendship with staunch royalist Viscount Brouncker. In 1661 the king granted the society a royal charter, which created the Royal Society of London.

The members of the Royal Society reinvigorated a scientific world outlook that had lain dormant for centuries. As methods for seeking objective meaning about the natural world began to diffuse widely in the seventeenth century, these men challenged one another to question traditional thought and to seek answers through reproducible, documented experimentation. Men who alone might have been an obscure cleric here or a university mathematician there pushed each other to stretch the limits of knowledge however they could. As Thomas Sprat described in his *History of the Royal Society*, "they had no Rules nor Method fix'd: their Intention was more to communicate to each other their Discoveries, which they could make in so narrow a compass, than an united, constant, or regular Inquisition."[9]

The result was the birth of a new intellectual age, a scientific revolution capped by the immediate recognition of the significance of the work embodied in Isaac Newton's *Philosophiae Naturalis Principia Mathematica* (*Mathematical Principles of Natural Philosophy*) published in 1687 under the imprimatur of Samuel Pepys, who was then president of the Royal Society of London. With the publication of the *Principia,* the heavenly bodies, viewed by Aristotle as divine, were brought into the range of human inquiry and shown to obey discernible laws of mathematics.[10] More broadly, as a later historian wrote, "The new learning, for long blocked by the Aristotelians, had by this time found its way into some of the Universities. The number of those concerned with natural philosophy was increasing rapidly."[11]

The discoveries of this era remain of epic significance in the history of science, as Herbert Butterfield, among others, has chronicled.[12] Less widely recognized, however, are the significance of the society's commitment to openness, its emphasis on recording and disseminating scientific findings, and its contributions to scientific communication. These social innovations were just as groundbreaking as the scientific method they supported. In contrast to the secretive alchemists of the Middle Ages, the Royal Society operated in the open. Its members corresponded avidly with experimentalists in any part of the world where sympathetic fellows could be found (although they seem to have been unaware of science in China, which was well established at the time). In particular, the first secretary of the Royal Society, Henry Oldenburg (a native of Bremen, Germany, living in London) took on the responsibility of, in his words, "entertaining a commerce in all parts of the world with the most philosophical and curious persons to be found everywhere."[13]

On the European continent, Oldenburg's correspondents included Christiaan Huygens, a Dutchman who published his own work on dynamics; René Descartes, a Frenchman living in Holland who suggested in his writings that unverified assumptions lay beneath the received wisdom of the scholastics (medieval philosophers who drew their inspiration from Aristotle); and Gottfried von Leibniz, a German who, working independently, invented calculus at around the same time as Isaac Newton. Correspondence was also established with members of similar societies in Italy and France.[14] As Sprat noted of the Royal Society's members, "they have begun to settle a Correspondence through all Countries; and have taken such Order, that in short Time there is scarce a Ship come up the Thames, that does not make some return of their Experiments, as well as of Merchandize."[15]

As this emphasis on the far-flung exchange of ideas suggests, early modern science was universal in several senses. The virtuosi were mainly educated gentlemen who found patrons for their work or were wealthy enough to fund their own inquiries, correspondence, and participation in scientific meetings. As a result, their research was not limited by the need for government support.[16] Most of the era's pamphlets and letters, along with the rare book, were written in Latin, making the results of experimentation accessible to educated individuals in a wide range of countries. And most thinkers saw their work in very broad terms as part of a common effort to understand nature, rather than an inquiry in a particular field. Early modern science, then, was subject to very few institutional, political, or disciplinary claims.[17] These conditions persisted in some measure into the eighteenth century, but as science advanced, the social and political context inevitably began to change.

From the Invisible College to Scientific Nationalism

By the mid-nineteenth century, the growth of the scientific community and its increasing specialization had given a professional character to scientific research in Europe.[18] As Donald deB. Beaver and Richard Rosen observe, a professional class of scientists emerged as part of a "dynamic organizational process which led to a revolutionary restructuring of what had been a loose group of amateur scientists into a scientific community."[19]

They traced the basis for this process to "the scientific community's ability to lay claim to support from the outside society and the society's ability to provide it."[20]

Individuals trained in experimentation became employees of publicly funded laboratories, and the term "scientist" began to come into usage. Increasingly scientists focused on problems with practical applications, such as the development of vaccines and advances in materials handling.

At the same time, science became increasingly nationalized. Government involvement in activities supporting science is nothing new. In England, the government's role in the standardization of weights and measures predated the Magna Carta, and many far more ancient civilizations relied on standard measures as well. Earlier governments also helped spur innovation by granting patents—temporary monopolies—on new discoveries. The Republic of Venice set down some of the fundamentals of patent law in 1474. A century and a half later, the British Statute of Monopolies limited the granting of patents to new inventions and set the life of a patent at fourteen years.[21] And in 1663, a second royal charter gave the Royal Society of London the right to publish the latest in scientific knowledge.

Patents and trademarks, a system of weights and measures, and trade rules were all developed to control and manage the fruits of creative activities. Government involvement in these activities does not predate science, but it was a prerequisite for the growth of professional science and the retention of knowledge within a national system. It also played a critical role in creating an enforceable legal framework that could encourage risk taking and entrepreneurial behavior, as economist Joseph Schumpeter pointed out.[22]

Public authorities, then, have long played a role in creating a framework within which science can thrive. But direct support for science is a more recent phenomenon. It developed in part because science became increasingly expensive—too much for any one patron or private group to support. As the sciences grew into professions, laboratories moved out of scientists' homes into universities and special institutes. These institutions petitioned governments for funds to support their research. In the nineteenth century, the French government became the first to use public funds to support science in dedicated laboratories and museums. (Some Italian cities may have offered funding through scientific societies during the eighteenth century, but not in a sustained way, nor to professional laboratories.) Paradoxically, England fell behind other countries in organizing public financial support for science and in recognizing science as a profession.[23]

In the United States, the Lewis and Clark expedition is often considered the first official example of the federal government's funding a scientific endeavor. President Thomas Jefferson approved the expedition and personally added a broad scientific mission to the original expedition to map a passage west. But the U.S. government did not make an institutional commitment to invest in science until 1863, when Congress chartered the National Academy of Sciences in the Morrel Act. The same legislation also instituted the system of land-grant colleges, a first step in developing what would become the system of public research universities in the United States.

As higher education expanded and scientific and engineering professions acquired increasing prestige, more people chose to go into the sciences. As a result, by the twentieth century the invisible college was not only more closely tied to national identity than in Boyle's day, but it was also much larger and more professional. In every respect, science in the twentieth century expanded at an unprecedented rate. The numbers of trained scientists, institutions, and budgets in wealthy countries grew exponentially. Derek de Solla Price observes that, by the late 1950s, "using any reasonable definition of scientist, we can say that 80 to 90 percent of all scientists that have ever lived are alive now. . . . The large-scale character of modern science, new and shining and all-powerful, is so apparent that the happy term 'Big Science' has been coined to describe it."[24]

From Big Science to the New Invisible College

How big was "big science"? In eight decades—from 1923 to 2005—U.S. government funding for research and development (R&D; a subset of science and technology) increased exponentially from less than $15 million to $132 billion per year (in constant dollars).[25] By the end of the twentieth century, R&D spending averaged 2.2 percent of gross domestic product among countries belonging to the "rich man's club," the Organization for Economic Cooperation and Development (OECD).[26] Total world spending on R&D reached $729 billion in 2000.[27] This number does not fully reflect spending on equipment and capital investments in scientific capacity, which are often budgeted separately from R&D, nor does it include activities like data collection or sample maintenance that are not considered "active research." These related science and technology activities constitute at least 20 percent more spending above R&D budgets.[28] It is possible, then, to argue that by 2000, global public spending on all the activities that can be considered within the purview of science and technology topped $1 trillion.

The emergence of big science was tied directly to growing appreciation of science's contribution to national security, following the outbreak of two global conflicts in three decades. In particular, the experience of World War II—culminating in the dropping of atomic bombs over Hiroshima and Nagasaki—had a critical impact on state interest in supporting scientific and technological research and development. As Price pointed out, "since World War II we have been worried about questions of scientific manpower and literature, government spending, and military power in ways that seem quite different, not merely in scale, from all that went before."[29]

In addition, in the years just after World War II, it became clear that science and technology catalyzed economic innovation and growth, even if the links were difficult to trace. Many important technologies developed during the war (such as radar, penicillin, atomic energy, and computing) were found to have economic value in the commercial market after the war. This outcome encouraged the creation of a close relationship between the policy and scientific communities. As Vannevar Bush, a physicist who headed a key U.S. Department of Defense research laboratory during the war years, argued in an influential essay titled "Science: The Endless Frontier":

In 1939 millions of people were employed in industries which did not even exist at the close of the last war—radio, air conditioning, rayon and other synthetic fibers, and plastics are examples of the products of these industries. But these things do not mark the end of progress— they are but the beginning if we make full use of our scientific resources. New manufacturing industries can be started and many older industries greatly strengthened and expanded if we continue to study nature's laws and apply new knowledge to practical purposes.[30]

Over time, interest in such practical uses came to be the principal driver of investment in science. As Jacob Schmookler argued in 1966:

The demand for science (and, of course, engineering) is and for a long time has been derived largely from the demand for conventional economic goods. Without the expectation, increasingly confirmed by experience, of "useful" applications, those branches of science and engineering that have grown the most in modern times and have contributed most dramatically to technological change—electricity, electronics, chemistry and nucleonics—would have grown far less than they have.[31]

The twin desires to create new military and civilian technologies led governments to take a leading role in shaping the direction of postwar science in the United States, and slightly later in Europe and Japan, as these regions recovered from war. Large federal and regional agencies came into existence to manage the relationship between the political and scientific communities.[32] A secondary set of publicly funded programs disseminated the results of science for economic application, often at the regional and local levels— think of county extension agents.[33] In Japan, for example, these programs included a group of more than sixty Kohsetsushi centers—research institutes with the task of making science available to local industries.

The growth of this system can be viewed as a set of feedback loops among institutions and functions. As science evolved, these feedback loops formed, enabling information to flow among the government, industry, and university research sectors. Loet Leydesdorff and Henry Etzkowitz describe this as a "triple helix" of interacting institutional functions within the innovation system.[34] Institutions evolve in size, scope, and function as information flows within the system. The feedback among sectors and the resulting changes in institutions allowed science, technology, and state institutions to co-evolve into mutually helpful entities.

The big science model was spectacularly successful in building scientific capacity and nurturing economic growth for a few wealthy countries. The infrastructure and financial support governments provided helped science and technology-based sectors to grow and flourish. Industrial development benefited from these investments, as did such public missions as energy and defense. But countries unable to make such investments fell further behind. From 1913 until the 1970s, the ratio of per capita income in the most developed country to the least developed country grew from 10 to 29. Many economists have attributed this growing gap to unequal access to science.[35] According to World Bank economists, investments in activities like education and science appear to have as large an impact on growth as physical capital investment.[36] Consequently, as economies increasingly became knowledge-based, the failure—or inability—of some countries to invest in science began to exact a heavy price.

Today the mechanisms of national science and technology policies remain the most visible parts of the system that produces science around the globe. The continued prominence of these mechanisms, however, belies the important changes that have taken place in the organization of scientific activity. Since the 1990s, the role of national policies in directing scientific research has diminished significantly; the influence of global networks, though, has grown. Multiple factors lie behind the growing influence of networks, including the rise of worldwide scientific capacity, improved access to communications technologies, and cheaper travel. The most important factor appears to be within the social network.

The transition from national systems to networked science began to gather steam around the same time as a number of other seismic events: the end of the cold war, the emergence of a unified Europe, the spectacular rise of electronic and digital communications, and the globalization of business. All these developments contributed to a sense that the world was changing in fundamental ways—that it was flattening, in Thomas Friedman's words. When it comes to the distribution of scientific activity, though, today's world is far from flat. The topography of science looks more like a series of mountain peaks towering over a flat plain. But the ability to access knowledge, contribute to it, and reap its benefits is distributed more broadly than ever before. The ability to link into science is still not evenly distributed, however—it has grown even more rapidly in countries that have traditionally made up science's core than in those that lie along the periphery. But the rise of networks has opened up the structure of science, creating new and more viable opportunities for poorer nations to participate in the system. Instead of taking the twentieth-century route of trying to replicate the national innovation systems of the United States, Europe, or Japan—an expensive and often fruitless enterprise—policymakers in developing countries should focus on understanding the forces that drive emergent, self-organizing knowledge networks. Based on this understanding, they can then create effective, feasible strategies for promoting their own membership in science's invisible college.

What Is the New Invisible College?

The people and the communications that constitute global science compose the invisible college. It is important to understand what types of communications make up the invisible college. As figure 2-1 shows, the partnerships that form the heart of the network fall into three broad types. Megascience projects like the International Space Station or CERN (the European Organization for Nuclear Research) form the tip of the iceberg in two ways: they are among the most visible forms of collaboration and they are relatively rare. These partnerships, which David Smith and Sylvan Katz call "corporate" (in the sense of "formal") are usually initiated to reach a specific goal.[37] Such activities, which are typically quite costly and are conducted over the long term, represent a small percentage of all international scientific endeavors when viewed in terms of overall spending. For example, the U.S. Congress Office of Technology Assessment (OTA) estimated in a 1995 study that, within the U.S. public research budget, megascience projects accounted for about 10 percent of the federal (defense and nondefense) R&D budget during the late 1980s and early 1990s.[38]

Many megascience projects take place at a centralized facility and require highly specialized equipment. The capital costs for such projects tend to be very high, requiring large-scale national and international funding. Government officials typically plan such facilities in discussion with scientists and sink significant investment in their construction before any research ever takes place. The organization of these activities can therefore be considered "top-down."

In contrast, the research projects that make up the base of the pyramid are "bottom-up," meaning that they are driven and organized collectively by individuals. For example, two or more researchers from different institutions might form a team to write an article, conduct a workshop, develop a database, or train a postdoctoral fellow. The goal in each case is generally to solve a problem or address a research question by sharing complementary capabilities. The participants need not spend all, most, or even any of their time in the

Figure 2-1. *Forms of International Scientific Collaboration*

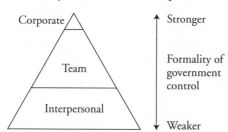

same place, nor will any single project necessarily absorb all of their attention. These projects are often short, lasting from one to three years, and each partner typically funds his or her own participation. Such person-to-person collaborations make up the vast majority of international collaborative activities.

Interpersonal collaborations are less visible than other forms of global science. Funding agencies have little control over how these project-level funds are spent and even less control over who participates in the research. Although it is relatively easy to track money going to megascience projects like the International Space Station, it is much harder to track spending on hundreds of informal collaborations in fields like soil science. In these cases, collaboration emerges from the requirements of the research and is chosen as the preferred method of research because it contributes directly to problem solving. The researchers themselves establish the collaborative links through professional networks. These links are the invisible base on which most of the global scientific community operates.

Between the tip and the base of the iceberg lie "team collaborations," which are formal projects involving the collective use of resources. In these cases, funds are usually granted to a principal investigator who has discretion to disburse the funds to other project participants. Team collaborations thus involve more centralized organization than person-to-person partnerships. They do not, however, approach the scale of megascience projects, and at least partly in consequence, government officials or other institutional representatives play a much less direct role in guiding or organizing such activities.

To gain further purchase on the activities that constitute the invisible college, we can also juxtapose the way in which research is organized (top-down or bottom-up) against the way in which it is conducted and typically located (centralized or distributed). Centralized research activities need to take place in a particular location, often because they require access to a particular resource or the facilities of a particular institution. Distributed research activities can essentially take place anywhere and everywhere at once, with a project leader or leaders dividing the research tasks among a large group with the intent of integrating findings at some later point. Four types of research emerge from this categorization:

—**Megascience** projects are generally well-known and highly visible examples of "heavy" research, such as the International Space Station or the multinational Large Hadron Collider, a facility for international fusion research located in Switzerland. These activities are highly centralized, organized from the top down, and frequently focused on a specific research goal.[39] Government officials often negotiate financial contributions and missions for megascience projects, which consequently are likely to serve both political and scientific interests.

—**Geotic** activities require sharing a resource at a specific place. Researchers travel to these sites or send data to be analyzed in unique conditions. For example, many different governments and research agencies share the international research center at the South Pole. And because conditions in a rain forest cannot be replicated in Europe, researchers must travel to the forest to conduct experiments. Geotic projects have centralized features—activities are organized in a corporate structure at a specific place where a manager may coordinate them—but they also have bottom-up features in that scientists determine individual projects.

—**Participatory** projects are centrally planned from the top down but are carried out in many places. Such distributed research undertakings are more difficult to identify and therefore more difficult to manage. The Human Genome Project, for example, involved six countries and dozens of laboratories in a highly distributed research enterprise. This project was planned by a leadership team, but executed by many people around the globe. The dynamics of distributed projects tend to resemble those of networks—there may be no clear leader, membership is voluntary, and the results of projects can be widely shared and highly diffused. (The Human Genome Project participants posted research results on the Internet at the end of each day.)

—**Coordinated** projects are initiated by scientists and take place in various widely distributed locations and laboratories. The Global Biodiversity Information Facility (GBIF) is an example of a bottom-up, highly distributed research activity. It involves dozens of countries and input from hundreds of researchers in a highly distributed research and database development venture. The GBIF website draws together these distributed activities, and GBIF results are freely available on the Internet.[40]

Taking the principle of distributed collaboration further, some research fields, such as seismology, have no central point of coordination. Researchers share data and information at conferences and with other scientists who are interested in their work. International activities in such disciplines are the most networked and the least organized.

Today all research is becoming more interconnected, collaborative, and networked.[41] Nevertheless, my research shows that, as a share of all collaborative investigations, distributed activities appear to be growing faster than centralized projects. Such coordinated and participatory activities present particularly great challenges to policymakers: If research is geographically distributed, how can knowledge be integrated into a useful whole? How can it be made available at the local level? Can tasks and resources be divided in a way that is both efficient and beneficial to all parties? These are among the questions considered in this book.

Who Funds the New Invisible College?

Table 2-1 shows different funding sources for global scientific collaboration. Governments commit more money than private sources, both as direct and indirect allocations. Private sector groups appear to be supporting international collaboration to an increasing degree, but their financial commitments are more difficult to track because they are not accountable to taxpayers. Some aspects of private sector investment in international collaboration can be seen, however, in the increased number of registered or announced corporate research alliances.[42]

National governments are a large and influential source of funding for global science but there is an important distinction between budgeted and actual spending by governments. In very few cases do governments intentionally fund international science. Governments fund science to promote national goals, as might be expected given that they are spending taxpayer money. These goals can be explicit and direct, such as the promotion of solar energy, or implicit and indirect, such as support for the knowledge pool needed for economic growth. Given their focus on national priorities and their accountability to the public, governments are loath to dedicate budgets to international collaboration unless they can argue that clear efficiencies will be gained. Explicit funding for international collaboration is perhaps between 5 and 15 percent of all national research and development funds in scientifically advanced countries.[43] These funds are typically committed to megascience projects or international organizations, such as the Human

Table 2-1. *Sources of Funding for International Collaborative Research*

Funding source	Example	Purpose of mission
Government agencies, institutes, public universities, special programs	National Science Foundation; Fraunhofer Institute, CNRS, Swedish Institute for Development Assistance	National defense, foreign relations, capacity building, innovation encouragement
Quasi-governmental bodies	World Bank (CGIAR), World Health Organization, NATO	Capacity building, poverty relief, public health, food production
Nongovernmental organizations	Rockefeller Foundation	Capacity building, poverty relief, public health, food production
Private companies	Cisco Systems, IBM, Siemens	Innovation, access to markets, cost reduction

CGIAR = Consultative Group on International Agricultural Research; CNRS = Centre National de la Recherche Scientifique.

Frontier Science Program, which devotes much of its budget to international collaboration on basic research in the life sciences.[44]

International organizations, including those such as NATO that are not explicitly scientific in focus, can allocate a larger percentage of their resources to international research activities. (In the case of NATO, this process takes place through its Science for Peace and Security Program.) But the overall amount of spending by international agencies is very low compared with government and corporate spending. Nongovernmental organizations commit considerable funds to science and technology-related projects, but they are often highly mission-specific (such as crop research or the development of a malaria vaccine) and do not always explicitly target international collaboration. In this way, the thin slice of the budget pie dedicated to international activities is quickly exhausted, but this is only the beginning of the story.

Agencies also unwittingly fund and influence the organization of collaboration by underwriting projects in which international cooperation is not part of the goal, but merely a means to an end. Team collaborations, for example, are not usually funded because they are international; they are funded because they are good science. Projects sponsored by the Consultative Group on International Agricultural Research (CGIAR), an intergovernmental group hosted by the World Bank, fall in this category, as do many of the projects funded through government agencies such as the U.S. NSF or the

Japanese Ministry of Education, Science and Culture. Similarly, person-to-person collaborations often draw on government funding, even though they may not have been anticipated, let alone mentioned, when a researcher's grant application was submitted or approved.

Any effort to tally the funds dedicated to international collaboration in science, then, is highly misleading. The megascience funds are a small portion of any public budget, and ironically, they often fund national science. Conversely, national science funding often funds a full range of international collaboration that is invisible to budget readers. Estimates of the amounts of public spending allocated to global science cannot be made based on budgets. To find this, we must look at a completely different set of indicators, which are the subject of chapter 3.

What's New about the New Invisible College?

In what sense is all this new? After all, researchers of different nationalities have always collaborated. And governments have long supported such efforts, whether to promote scientific or foreign policy goals or both. Consider the Manhattan Project, perhaps the defining megascience project of the twentieth century. Sequestered in Los Alamos, New Mexico, émigrés from war-torn continental Europe, such as Hans Bethe and Edward Teller, toiled alongside American scientists to build the world's first atomic bomb.

Today such government-directed international collaboration continues. But what is new about the new invisible college is the shifting balance of power between "international" and "global" science resulting from the flow of communications beyond borders. Despite their surface similarity, international and global science concepts are quite distinct. International science describes activities in which people are working in more than one country or receive their equipment and funding from multiple countries, or both. The term implies that collaboration occurs fundamentally between nation-states and that groups of researchers from these nations work together with the support and protection of their governments. This approach, which dovetailed with the ideology of scientific nationalism, characterized science in the twentieth century—a period, not surprisingly, when megascience projects dominated the international collaboration agenda and political treaties tried to promise entry into scientific collaborations.

Global science, by contrast, describes activities in which researchers are free to join forces to tackle common problems, regardless of where they are geographically based. Global science is growing not because nations are promoting it, but because it serves the needs of those working within the knowledge-creation system. In this way, it shares some features with the globalization of business, which also extends beyond the interests of the nation-state. But unlike business, the growth of global science is not driven primarily by financial needs. The invisible college is driven by the needs of the knowledge-creating community, which in turn is driven by the desire to do original and creative research. In the next chapter, I discuss these motives in greater depth and show how they shape the rate and direction of the invisible college's growth.

CHAPTER THREE

NETWORKED SCIENCE

> It is surprising that such a simple law should be followed so accurately and that one should find the same distribution of scientific productivity in the early volumes of the Royal Society as in data from the twentieth-century Chemical Abstracts . . . the regularity tells us something about the scores we are keeping.
>
> DEREK DE SOLLA PRICE, *Little Science, Big Science*

The forces driving the emergence of the new invisible college can be discerned and put to work to improve the productivity and distribution of scientific activity. This chapter approaches this task by drawing on recent work in a number of disciplines, including political science, sociology, mathematics, and computer science. Joined together, new methods of studying the dynamics of social systems can reveal the structure of the global network, which in turn can be shown to follow predictable mathematical probabilities and social "laws." But like the invisible college itself, the basic concepts that lie behind the reorganization of science are not new. Although the seventeenth-century virtuosi would be bewildered (or perhaps inspired) by the advances computer science has made possible in such fields as systems theory, complexity theory, and cybernetics, they intuitively grasped the importance of understanding complex systems as irreducible wholes. In fact, one of the first works commissioned by the Royal Society of London emphasized just this point.

Derek de Solla Price, *Little Science, Big Science* (Columbia University Press, 1963), p. 43.

Seeing the Trees and the Forest

Among the first official requests to the newly established Royal Society for a scientific study came one from the Royal Navy. Although the Royal Society was populated by men who studied the heavens—and this might have been a likely subject of naval interest—the Royal Navy's inquiry was about trees. The Commissioners of the Royal Navy were concerned about the availability of wood. A single gunship required up to 3,800 trees, which equated to about 75 acres of forest. Wood was a critical and diminishing natural resource in seventeenth-century England, and information on the revival and sustainability of forests was viewed as critical to national security. Deforestation had become a significant problem, with social implications that extended well beyond the interests of the Royal Navy.

The task of studying the management of trees and forests was assigned to Royal Society fellow and able horticulturalist Sir John Evelyn. Drawing on years of his own research as a landed gentleman, he produced a compendium of information about indigenous trees so thorough and comprehensive that forever afterward he was called "Sylva" after his book: *Sylva, or A Discourse of Forest-Trees and the Propagation of Timber.* Published in 1664, *Sylva* was one of the first books commissioned by William, Viscount Brouncker, the Royal Society's first president.

Every seventeenth-century British subject knew that an oak tree grew from an acorn, so why commission a study on this subject? The problem arose not in knowing how a tree germinates, but in understanding the interactions among different plants and trees and other parts of the environment. These interactions determine the health of individual trees and the well-being of the forest as a whole. Evelyn knew this intuitively from his years of systematically studying horticulture. In *Sylva,* he discussed the forest as an ecosystem—without using the term, which was not introduced until the twentieth century. He described the system's components (soil, water, plants, and animals) and the emergent order of which they were a part (the forest). He showed how these physical and biological components related to each other and how they interacted with an environment that included both cooperative and competitive features.

In short, *Sylva* established that a forest is a complex adaptive system. It is complex because it is composed of many different interacting elements. It is adaptive because it can change in response to shifts in the environment, including changes in any of the constituent parts. And it is a system because it is a collection of organized objects that form a recognizable whole. Like most complex adaptive systems, a forest is also open—new elements can move across its boundaries and become incorporated into the system, and existing parts can exit without causing the system to collapse. And perhaps most important for our purposes, a forest is *emergent.*

An emergent system takes form on its own, rather than being planned or established by edict. The organization of such systems is not determined by a government or a corporation or a blueprint. Once organized as a recognizable unit, an emergent system forms an irreducible whole whose form and function exceed the sum of its parts. For example, if we were to mix together the elements that make up a tree—water, gases, minerals, and so on—no amount of wishing, waiting, or hoping would create a tree, let alone a forest. The tree cannot be constructed from its component parts; nor can it be reduced to those parts and still be a tree. The same can be said of the invisible college.

The Invisible College as a Complex Adaptive System

Like a forest, the invisible college is a complex adaptive system. It is complex: millions of researchers around the world interact in both competitive and cooperative ways, with no overall direction. It is adaptive: both the college and its constituent researchers respond to changing environmental conditions such as shifts in the priorities of grant-making organizations or new discoveries. As Robert Axelrod points out, such adaptation "may be at the individual level through learning, or it may be at the population level through differential survival and reproduction of the more successful individuals."[1] In other words, individual scientists may choose to pursue new questions, or those who prosper under new conditions may end up training more numerous and more talented students who carry on their work.

The invisible college is recognizably a system, a collection of individuals and institutions devoted to the common pursuit of scientific knowledge. And it is open: scientists can cross from field to field or open up brand new areas of study. The development of biochemistry in the early twentieth century serves as an example. The two separate fields of chemistry and biology found increasing common ground on which to integrate themes and research goals. As their interactions grew more complex and became institutionalized, the new subfield of biochemistry emerged. Today, researchers find that biochemistry is the most interdisciplinary of all scientific fields.[2] The nascent subfield of nanoscience is now going through a similar process. Although not yet fully defined, it is integrating and coalescing out of physics, chemistry, materials science, and biology.[3]

Finally, as the rise of new fields suggests, the invisible college is emergent. No central agency dictates its organization and growth. Instead, its direction is shaped by interactions among individual scientists who communicate with each other to share findings and puzzle over results, form partnerships when needed, and change course in response to new opportunities and constraints. By deciding what type of research to pursue and with whom, along with when, where, and how to conduct that research, they collectively determine how scientific activity is shaped and how knowledge evolves.

Why does characterizing the invisible college as a complex adaptive system matter? Simply because of this: physicists have found ways to measure the dynamics of these systems and to predict—at least probabilistically—how they will evolve. Even though activity within the invisible college is largely self-directed, it is not random. It follows identifiable patterns and rules. By uncovering these patterns and rules, we can understand not only how the invisible college works, but also how policymakers can influence its evolution and growth and how its benefits can be distributed.

The first step toward this goal is recognizing that the invisible college is a particular kind of complex adaptive system—a "scale-free network." To understand what this means, we need to investigate three questions: What is a network? What is a scale-free network? How do such networks operate?

What Is a Network?

A network is a formal way of describing any set of interconnected relationships among actors or things. Networks are constructed from components that stand alone but can be made interdependent. Think of the parts of a transportation network, such as an airport, a bus route, and a highway. Each is a subsystem of the larger transportation system, each can stand alone, and all three can be interconnected to create new efficiencies and capabilities. Similarly, within the invisible college, research groups, disciplinary fields, and institutions can stand alone, but by creating connections with other elements of the system, they can greatly enhance their value.

The invisible college has an additional feature that differentiates it from a transportation or other infrastructure network: it grows solely out of its members' interest in communicating with one another—in other words, it is an emergent system. These connections can be traced by examining the published literature to which they ultimately give rise.

The invisible college is particularly well suited for study as a social network because it offers a great deal of data about the relationships it nurtures. For example, it is possible to collect the names of scientists who participate in particular projects or who are employed at particular institutions and then trace the links among them.[4] The authors of scientific articles typically acknowledge intellectual debts to other researchers through citations, references, or text acknowledgments.[5] Further evidence can be collected on coauthorship patterns, including the extent to which scientists collaborate with researchers in other fields, institutions, and countries. These data help to reveal the knowledge networks that support published work.

The invisible college emerges from careful decisions on the part of researchers to share resources, particularly when collaboration involves a long-term commitment. In making such decisions, individuals weigh costs and benefits. The costs of forming a research partnership can be high. They are likely to take the form of time and funding spent on a project (and the opportunity cost of time and research dollars not spent on something else). In addition, collaboration requires researchers to give up some of the exclusivity surrounding their data or results (as demonstrated in the BeppoSAX project). To benefit from a nondirectional network, in which information flows in both directions between partners, participants must share valuable information (reciprocity) or provide a resource (complementarity). Participation, then, is not free. Although the network may be open to new members, a prospective new member must have something to share—such as experience or a resource—that makes that individual attractive to existing members.[6] Moreover, as a network matures, the price of entry rises. For example, a research team that is a year into its work is unlikely to welcome a newcomer unless that person brings transformative data or unique research capabilities.

What do scientists receive in return for their contribution to a research network or team? The immediate benefits often involve access to specific and unusual resources, complementary or integrative capabilities, well-connected people, and, of course, funding.[7] For example, scientists can tap into a network of colleagues to gain access to a scarce resource, such as a soil sample, or unique data, such as the position of gamma ray bursts. Direct benefits such as these typically motivate the formation of a particular partnership.

The indirect benefits of collaboration, though, are at least as important as the direct results. By participating in research partnerships, scientists strengthen their membership in the broader network that makes up the invisible college. And in so doing, they gain greater access to the global exchange of free-floating ideas and information within the weak links (or "weak ties," which are those friend-of-a-friend associations that can prove to be useful connections to information, as explained in detail later in this chapter) of the network— ideas and information that might ultimately take research in unexpected directions. Social networks such as the invisible college fulfill this function in part by offering informational shortcuts. Given the size of the research community, there is no way that one person can know everyone else who has potentially useful knowledge—or even who those people are. By consolidating such information and distributing it, whether through formal channels like directories, or through person-to-person connections, networks provide an essential service.

Even more important, social networks promote the sharing of knowledge and resources by creating trust.[8] By facilitating repeat interactions, the circulation of information (including, crucially, information on individual reputations), and the evolution of group norms, they enable the formation of what Francis Fukuyama calls communities with shared ethical values. Such communities, Fukuyama believes, "do not require extensive contract and legal regulation of their relations because prior moral consensus gives members of the group a basis for mutual trust."[9] Others have described this essential feature of networks as "social capital."[10] As the term implies, social capital is a commodity that a group can earn, bank, and spend. It is earned primarily by developing or learning and abiding by the norms of the community, speaking a common language, and exhibiting shared values. (Try speaking the language of physics with a physicist.) As Fukuyama points out, social capital cannot be acquired by individuals acting on their own—it is a property based on social attributes of the group rather than those held by the individual.

Collaboration appears to be more likely to emerge and then to be more fruitful within scientific networks with extensive social capital. The members of such networks are willing to share information more freely and make longer-term commitments to joint projects because they have greater confidence that these commitments will pay off and that reciprocity will be honored. The norms of reciprocity and fair play that characterize such networks (as well as the knowledge that violations of those norms are unlikely to be tolerated) help to allay scientists' fears that other researchers might steal their data, falsify results, or represent shared work as their own. By contrast, in sparse networks with little social capital, risks associated with sharing information may leave scientists to work alone—and less productively—with little in the way of feedback. (Think of the archetypal mad scientist.) The data suggest that scientists are increasingly seeking to work in these types of collaborative teaming arrangements, both creating and benefiting from the social capital that, in theory, allows the achievement of certain ends that would not be attainable in its absence.[11]

The invisible college facilitates the creation of social capital—and ulti-mately, the creation of knowledge—by organizing practitioners of science into self-identified and self-selected clusters organized at the level of scientific disciplines and subfields. The edges of a discipline are blurry (at what point in its practice does biochemistry differ from molecular biology?) but at some point for both insiders and outsiders it is possible to say that a person belongs to this scientific field and not that one. This categorization is based on a researcher's knowledge, training, contributions, and access to resources, as well as how that person defines what she does. In this way, it becomes possi-ble for a scientist such as Ulla Lundstrom, a renowned Swedish environmen-tal scientist, to stake claims in soil science research. This identification is sup-ported partly by the subjects she has studied and partly by the network of people who recognize her as a soil scientist and are willing to include her in the exchange of specialized communications.[12] These people create a social group that is identified by their mutual recognition, their topic of interest, and the level of trust that members have in the work of others—some of whom they know personally and some of whom are vetted through connec-tions to other trusted members.

If they are to survive, such groups must find some balance between stabil-ity and variation in their identification as a group. In science, each field or subfield develops a common nomenclature and set of techniques. By coalesc-ing around a pool of accepted principles and information, researchers can accelerate the acquisition of new knowledge. Each new experiment or discov-ery builds on a common framework, and no one is forced to reinvent the wheel. This stock of information can be thought of as "knowledge capital," or in Thomas Kuhn's terms, a "scientific paradigm."[13] Ultimately, however, the field cannot grow in the absence of variation in the form of new ideas and forms of expertise. Usually groups respond to this challenge by develop-ing specific norms for acknowledging, assessing, and accepting challenges to the prevailing paradigm.

What Is a Scale-Free Network?

To gain an understanding of some of the special properties of the new invisible college, it is helpful to visualize its structure using the language of network theory. The elements of a network are called nodes or points. A con-nection or communication between nodes is called a link: the links are visu-alized by drawing a line between points. Nodes with a large number of con-nections are called hubs. Figure 3-1 depicts these features, as well as some of

Figure 3-1. *Features of a Social Network*

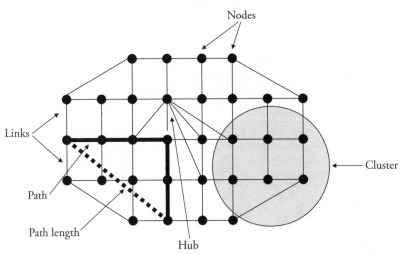

the other basic components of a network. For example, imagine the air trans-port system in the United States. Each airport, no matter how large or how small, is a node. The flights between them are links. Major city airports, such as Denver and Atlanta, which collect travelers and send them on their way, are the network's hubs.

The invisible college can be visualized in exactly the same way by repre-senting each researcher as a point and the connections among them (training, coauthorship, and so on) as links. Next, if we study the relative frequency of connections across researchers by constructing a chart in which each vertical bar represents the number of people who have a certain number of links, a very interesting pattern emerges. We might expect the distribution to take the familiar form of a bell curve, in which a few people have one or two con-nections, a few have thousands, and the rest fall somewhere in between. But in fact, the distribution of connections in the chart we have just constructed is scale-free, following what mathematicians call a "power law."[14]

A power-law distribution starts high along the vertical axis and decreases rapidly (figure 3-2). Contrast this to a bell-shaped curve, such as a normal distribution (figure 3-3). A normal distribution peaks at its average value: its average and modal (most common) values are the same. A power-law distri-bution peaks at its lowest value, and the average value lies somewhere to the

Figure 3-2. *Sample Power-Law Distribution*

Probability that any one node has that many links (percent)

Number of links

right of the peak. The average value, however, is in no way typical. For this reason, networks following this distribution are termed scale-free.[15]

Many analysts prefer to present power-law graphs in log-log terms. These graphs, which use logarithmic scales on both axes, transform the curve shown in figure 3-2 into a set of points that approximate a downward-sloping straight line. Power-law graphs are the signature graphs of a complex adaptive system.[16]

Scale-free distributions can be found in a wide array of settings, from the Internet to the interactions among proteins in a single cell.[17] In all these systems a few elements are extremely large or frequent or well connected, and the vast majority are very small or rare or essentially isolated. Consider, for example, the World Wide Web—although a few sites like Google garner the lion's share of traffic, most Web pages get no more than a couple of clicks a day.[18] The same holds true for the network of routers that powers the Web.[19] Similarly, the size of cities around the world follows a scale-free distribution, with a handful of giant cities followed by a few large cities, many smaller cities, and hundreds of thousands of small towns. The same distribution applies to wealth—there are a handful of billionaires and many more millionaires in the world, but the majority of people have little or no wealth. This skewed distribution is found in many systems.[20]

Figure 3-3. *Sample Normal Distribution*

Number of men of that height

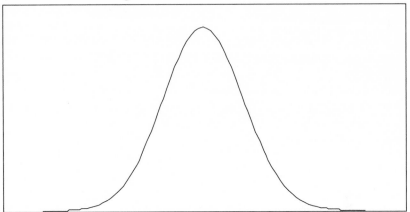

Height of adult males

Within science, most individuals are connected to only a small number of colleagues, usually at the same institution where they work or were trained. But a handful of star researchers—the Nobel prize winners or heads of major labs—have trained, supervised, or collaborated with thousands of scientists around the world. This pattern, which was established by Alfred Lotka as early as 1926, is reflected in the distribution of citations in academic literature.[21] Nearly forty years later, Derek de Solla Price showed that the number of scientific papers published by scientists followed Lotka's law (as referenced in the epigraph at the beginning of this chapter).[22] And more recently Mark Newman demonstrated that the networks of coauthorships formed by copublication of scientific papers at the global level also have a scale-free structure.[23]

These findings are important because if the invisible college is in fact a scale-free network, we can assume that it behaves in ways similar to other scale-free networks. For example, such networks are fairly resilient to accidental injury but can easily be paralyzed by the removal of a few major hubs.[24] More important for our purposes, the same mechanism appears to guide the growth of a wide range of scale-free networks. Physicists Barabási and Albert have dubbed this phenomenon "preferential attachment."[25]

The theory of preferential attachment describes how new entrants choose the actors with whom they want to connect when joining a network. These choices are typically constrained by the availability of connections and by the entrant's standing in the network, but by and large new network members try to connect with those who are better known and better connected. The appeal of well-known, well-connected individuals is obvious: they offer many benefits to fledging members such as newly minted Ph.D.s. Well-connected scientists control data, equipment, funding, and access to other resources and opportunities. As a result, they attract connections—and higher quality connections—at a far higher rate than less famous researchers. Ultimately, this process generates a scale-free structure, in which a few stars or hubs outshine the far larger number of ordinary researchers.[26] I explore this process and its implications in greater depth in chapter 4.

How Do Networks Operate?

Preferential attachment shows how the topography of a network like the invisible college evolves. To explain how such networks function and how the connections that constitute and support them are formed, though, we must look to other phenomena. Three concepts are central to this explanation: "weak ties," the "small-world phenomenon," and "redundancy."

Weak ties are those social connections that we tap infrequently.[27] An individual scientist's strong ties are likely to include his family, his close friends, and colleagues that he encounters daily. His weak ties, in contrast, might include a researcher that he sees from time to time at a conference, a college acquaintance with whom he has occasional contact, or a fellow commuter that he sometimes encounters while waiting for the train. When one visualizes a network of relationships, weak ties can be seen as relatively thin or unconnected spaces between nodes.[28] Nonetheless, they can be very valuable. Weak ties can link us, for example, to a "friend of a friend" who can offer us information, entertainment, or connections to others. To continue with the example of the soil scientist, her research project may require that she export soil samples (as many of these projects do). She may not know anyone who is familiar with export regulations, but she might mention this to a colleague visiting her town for a seminar, who knows someone she should call.

In his seminal work on this topic, Mark Granovetter found that weak ties play a crucial role in bridging clusters within social networks.[29] In science, for example, a researcher's strong professional ties—those people with whom he interacts daily or has very close bonds—tend to work in the same lab, institution, or field. If he wants to reach outside his own community and communicate with scientists in another discipline, a weak tie—say, a visiting speaker with related interests but a different set of connections—is more likely to form the crucial link, that variation needed to introduce new ideas.

Weak ties are thus more likely to grant access to people with different and challenging ideas or connections that reach beyond an individual's circle of daily contacts. The introduction of new ideas through weak ties can spur innovation and serendipity. In recognition of this connection, some institutions pay particular attention to fostering such links. The Center for Nanoscale Materials (CNM) at Argonne National Laboratory in the United States is one example. The emerging field of nanoscience thrives on weak ties among people drawn from many different disciplines. Consequently, the CNM has embraced a new organizational model built around flexible teams, as Derrick Mancini, a physicist at the center, explains:

> Due to the interdisciplinary nature of nanoscience research, an old academic model won't suffice. We try to encourage fluid and flexible teams—it's different from the old traditional model and different even from universities, which are intended to be a center to get funds. People come [to the CNM] with an idea of what they want to do . . . we have to realize that some people are preoccupied with instrumentation, others with technique, and others with the science problem. The best way to organize research is to reward people for doing what they do best. This works better with teams, where people can connect with others when they need to, and then go back to their bench to solve their problems.[30]

This is not to say that strong ties are unimportant. Strong ties help create social capital and build and retain knowledge within certain parameters, or, in other words, the stable base of Kuhnian "normal science" (which is the normal day-to-day practice of established scientific disciplines). But in a network composed predominantly of clusters bound by strong ties, with very few weak ties between them, there will be only limited communication and cross-fertilization of expertise across groups. Knowledge will be quite localized, whether by geography or by field. In contrast, by helping weak ties proliferate, the network and its members can create more fertile conditions for exchanging ideas and expanding knowledge. The CNM is organized to facilitate this exchange by making it easy to tap both strong and weak ties as needed.

Weak ties operate in part by strengthening the small-world phenomenon.[31] This is the common idea that every individual lies within six degrees of separation of any other person on the planet.[32] (In Hollywood, every individual lies within six degrees of separation of Kevin Bacon.)[33] Translated into network terms, this means that it is possible to link any two nodes within a large social network through a small number of steps.[34] Such a sequence of steps or links from node to node is called a path, and the distance between any two nodes within a network is called a path length.

Based on his study of scientific networks, Mark Newman concludes that the new invisible college is largely characterized by short path lengths:

> We find that typical distances between pairs of authors through the networks are small—the networks form a "small world" in the sense discussed by Milgram—and that they scale logarithmically with total number of authors in a network, in reasonable agreement with the predictions of random graph models. . . .We also show that for most authors the bulk of the paths between them and other scientists in the network go through just one or two of their collaborators.[35]

This last finding is yet more evidence of the importance of hubs in social networks, a pattern also noted by Milgram in his original work.

Short path lengths can be particularly valuable to low-status network members who lack many direct connections, particularly to hubs. For example, the average postdoctoral fellow is unlikely to receive a promising response if he calls one of the field's stars out of the blue and asks to collaborate on a project. But if, for example, his adviser went to school with someone who trained under that star researcher, that individual might be persuaded to make the critical call.

The small-world phenomenon clearly operates through strong as well as weak ties. Tightly integrated clusters within networks are small worlds by definition. Weak ties, though, greatly extend the reach of small worlds and create the potential for direct linkages among a far greater range of individuals or network elements. In this way, weak ties can give rise to particularly creative and fruitful forms of collaboration and exchange.

The small-world phenomenon is closely related to the notion of redundancy, or the existence of multiple paths between nodes in a social network.[36] Redundancy gives rise to clusters, groups within a network that are rich in links and consequently have short average path lengths. Redundancy can make a network highly resilient or robust; that is, its connectivity as a whole is not weakened by the random removal of links or nodes. Think of a transportation network in which there are redundant methods of traveling. If your car breaks down, you might be able to take a bus to work.

Scientific researchers almost always have many colleagues in common, or redundant social connections. Their common connections are held with those who share similar training and knowledge, who speak a common technical language, and who both cooperate and compete for recognition and resources. Redundancy promotes stability within the social network as well as retention of knowledge in both tacit and explicit forms. It also fosters productivity within the network by ensuring that valuable collaborations can be assembled in many different ways and that these collaborations do not depend on the availability of individual nodes or the strength of individual links.

The Royal Society as a Social Network

The history of the first invisible college can illustrate how some of these network features operate. Consider, for example, how Jan Amos Komensky came to be introduced to the members of the future Royal Society. Komensky, who is better known by the Latinized name *Comenius,* is often credited with having invented modern science-based education in Europe.[37] In the 1630s, when he was working as a minister and educator in Moravia (now part of the Czech Republic), his writings came to the attention of a man named Samuel Hartlib. Hartlib was to the seventeenth century what a Web portal might be to the twenty-first century. Born in Elbing, West Prussia (now part of Poland), and educated in Germany, he emigrated to London in the 1630s to escape the Thirty Years' War. His gift for languages led him to take on the role of an "intelligencer"—an agent for the dissemination of news, books, and manuscripts from throughout Europe in London. In this role, he met many members of the intellectual class.[38]

Hartlib was so impressed with Comenius's writings on universal education that in 1637, he personally arranged for Oxford University to publish the work.[39] He also prevailed on Comenius to make the arduous journey to England, where he arrived in September 1641 and remained until June 1642. The goal of Comenius's visit was to interest members of Parliament in his ideas on education.[40] Unfortunately the brewing conflict between the King and Parliament left little time for those engaged in government to talk about educational policy with a Moravian clergyman. Nonetheless, Comenius took advantage of his London stay, writing an important treatise on education and meeting with a small group of natural philosophers organized by Hartlib. Among these contacts was a previous acquaintance of Comenius, a German expatriate and co-religionist named Theodore Haak. Haak in turn was quite pleased to introduce Comenius to leading London intellectuals, such as Robert Boyle.[41] Some historians suggest that it was in the course of one of these gatherings that the idea of an invisible college of experimentalists was first broached, and that it was Comenius himself who suggested the term to the British experimentalists.[42]

This episode illustrates a number of features common to social networks. For example, Hartlib plays an important role in the story as a hub connecting educated Londoners not only to each other but also to intellectuals on the continent. Others among the virtuosi might have been able to make contact with Comenius, but they would not necessarily have had the connections that convinced him to cross half of Europe, as well as the Channel, to share his ideas. Once Comenius had made his hazardous journey, weak ties and the small-world phenomenon helped Hartlib in his efforts to introduce him to like-minded natural philosophers. Haak turned out to have a weak tie to Comenius, which made him more receptive to the idea of hosting a gathering. And that gathering seems to have been more significant than anyone expected because it gave rise to the idea that the participants constituted an invisible college. But if Comenius had not floated the idea on that evening, it is likely that it still would have taken hold eventually. The virtuosi, whose company he had just entered, were connected by many redundant ties, creating fertile ground for the circulation of ideas.

The Emerging Labyrinth

Today the knowledge system is an emergent network that can be thought of as a labyrinth with many possible paths.[43] Pathways through the knowledge network are created as preferential attachment, small worlds, and weak ties, which connect people to each other and to the resources they need to innovate. The resulting network emerges from the interests of individuals and can no more be deliberately constructed or created from its component parts than can the oak tree.

Similarly we cannot anticipate the scale and scope of scientific discovery, particularly when it comes to the fundamental paradigm shifts that make up scientific revolutions. Most advances in science and technology are evolutionary, constituting normal science in Kuhn's terms.[44] Normal science advances by recombining existing knowledge.[45] These advances are incremental and thus relatively easy to anticipate. But truly new discoveries— those that change the path of scientific progress—are nearly impossible to predict.[46] They often result from the combination of insights from very different fields, as Bob Hwang, a physicist who heads the Center for Functional Nanomaterials at the U.S. Department of Energy's Brookhaven National Laboratory, explains:

Our present definition of nanoscience comes from the fact that materials scientists, physicists, and others have recognized that something fundamentally different happens to materials at the nanoscale—different from what you see at other levels. We [scientists] discovered this phenomenon—so in that sense it grew out of the traditional work in science, but in another sense it is new: new departments in universities are forming all the time because of that particular discovery. The concepts underlying nanoscience are so encompassing that they stretch over all the present disciplines. . . . The real success in science comes when one takes a multidisciplinary approach, because this [discovery] would not have happened otherwise.[47]

Scientific knowledge emerges from the combination of people, ideas, and resources.[48] Emerging ideas—such as the theories offered for the different structure of materials at the nanoscale—compete for attention within communities of researchers, just as fledgling oak trees compete for resources on the forest floor. Good ideas are discussed, codified, peer-reviewed, and published in journals or as patents or standards. The best or most widely accepted ideas stabilize into a paradigm and are used again and again, as evidenced by citations within journal articles, licensing of patents from one company to another, or marketing of products. Some ideas—think of gravity—become so widely accepted that their originator is no longer cited.

Given the difficulty of anticipating the location, scale, and scope of new discoveries and developments, research cannot be designed in advance: only the conditions and incentives to encourage it can be set in place. These conditions create the landscape in which science can thrive. This landscape consists not only of a critical mass of people, infrastructure, and institutions, but also of the networks that connect them. Unfortunately, these networks can be difficult to see, and efforts to tap into them are daunting for policymakers and scientists who do not know their structure, speak their language, or understand how they work. To help make these networks more visible and accessible to those who wish to join or support them, I detail new ways to both understand and govern the emerging knowledge labyrinth in the chapters that follow.

PART **II**

THE LABYRINTH OF THE WORLD: UNDERSTANDING NETWORK DYNAMICS

"But where is your guide?"

"I have none; I trust God and my eyes not to lead me astray," I answered.

"You will accomplish nothing," he replied;

"Have you ever heard of the Cretan labyrinth?"

"Yes, a little," I assented.

"It was one of the wonders of the world," he continued; "a building with so many rooms, partitions, and passages that anyone entering it without a guide was doomed to wander and grope about it without ever finding his way out. That, however, was a mere joke in comparison with the arrangement of the labyrinth of this world, especially in our day. Take the advice of an experienced man and do not trust yourself into it alone!"

JAN AMOS COMENIUS
The Labyrinth of the World and Paradise of the Heart, 1657

CHAPTER FOUR

TECTONIC SHIFTS:
THE RISE OF GLOBAL NETWORKS

> Surprised though we may be to find it so, the scientific world is no different
> now from what it has always been since the seventeenth century. Science has
> always been modern; it has always been exploding into the population,
> always on the brink of its expansive revolution. Scientists have always felt
> themselves to be awash in a sea of scientific literature that augments in each
> decade as much as in all times before.
>
> DEREK DE SOLLA PRICE, *Little Science, Big Science*

T he seemingly solid ground beneath our feet sits on shifting tectonic
plates. In human terms, the plates seem to move slowly, but across the
millennia, their movement has created the continents and oceans as we know
them today. The shifting of tectonic plates gives rise to earthquakes, which
alter the physical landscape that rests on the plates. These shifts are driven in
turn by forces emanating from deep within the earth. According to the theory of plate tectonics, which was proposed independently by Harry Hess and
Ronald Dietz in the early 1960s, the surface of the ocean floor expands as
magma pushes up from inside the earth.[1] As magma emerges, the plates on
which the earth's surface sits are pushed away from the newly created surface,
and the older parts of the plates' edges are pushed down.

Like the physical landscape around us, the social landscape of scientific
inquiry changes in response to shifts in its underlying structure. During the
past 350 years, this underlying structure has shifted from an individual-based
system to a professional one, through a nation-based system, and finally to the
network-based system of today. Science still takes place at laboratory benches
and field locations around the globe, but the communication structures that

Derek de Solla Price, *Little Science, Big Science* (Columbia University Press, 1963), p. 11.

help drive advances in science and technology no longer rely primarily on national institutions. Instead, scientific networks operate and interconnect practitioners locally, regionally, and globally, with little regard for national borders.

This shift in the organization of science results in part from the speed of transactions offered by advances in information technology since the early 1990s. The spread of the Internet and the invention of the World Wide Web catalyzed a shift that was already under way. These developments enhanced the ability of scientists and nonscientists alike to access and share scientific tools and knowledge. By lowering transaction costs, they also increased the productivity and efficiency of research, notably by facilitating distributed collaboration. But these technologies did not cause the change in the way science is organized, any more than the San Andreas Fault caused the 1994 Northridge earthquake in Los Angeles. Although seismic faults demonstrate that tectonic plates move, they do not cause those movements.

Like the earth's tectonic plates, the organization and conduct of science respond to bottom-up forces—in this case, the motivating forces that drive individual scientists to communicate with each other. These forces are altering the landscape for scientific inquiry, as well as for funding and policy in the twenty-first century. In this chapter, I focus on identifying the factors that motivate collaboration and cooperation among scientists at the global level. I also show how these forces are reflected in the growth of the new invisible college and its evolution over time.

Looking inside the Earth

The story of Michael Fehler and Haruo Sato helps to illustrate the forces that promote collaboration within the new invisible college. No agency, institution, or organization introduced the two geophysicists—one hailing from the United States, the other from Japan. No global ministry of science and technology organized them into a team. Instead, the two scientists became friends in 1984 when Fehler, then employed by the U.S. Department of Energy's Los Alamos National Laboratory in New Mexico, visited Japan to conduct research on geothermal energy. During his visit, Fehler contacted Sato on the recommendation of his former dissertation adviser at MIT. Sato proved to be not only a gracious host but also an inspiring colleague. Soon the two scientists found that they were interested in similar research questions about seismic waves—whether the waves were created by earthquakes or by underground nuclear explosions.

Their collaboration flourished in 1988 when Sato obtained funds to invite Fehler for an extended stay in Japan. The two took advantage of this opportunity to write a number of papers together as well as to outline a book on seismic wave propagation and scattering, which was published by the American Institute of Physics in 1997.[2] Their work was quickly cited by dozens of other scholars, making the partners attractive hubs in the invisible college of seismology research.

Fehler and Sato's collaboration flourished not only because of their complementary capabilities, but also because of their similar drives for creativity and discovery. Each partner did high-quality, innovative research; each brought new ideas to their joint work; and each had access to a different funding stream—Fehler in the United States and Sato in Japan. Equally important, the two scientists discovered that their personalities meshed. As Fehler explained, "There are many people who are doing this research now, but there are a few of us that are pushing the envelope of what can be done. We are interested in the same science, but we also get along really well; there is a lot of respect and trust."[3] Together, these advantages more than compensated for the differences in time, place, language, and culture that the two researchers had to overcome in order to work together.

In 2005, Fehler and Sato met at a conference in Santiago, Chile, eager to discuss the most exciting advance in seismology in a long time. Their discussion was about noise—not the typical conference chatter, but the use of noise as a new measurement tool in seismology. Seismic events generate waveform data that are captured by the Global Seismographic Network (GSN), a collection of 128 seismographic stations based in 80 countries and located on every continent. Scientists can analyze these data—notably the time it takes for waveforms to travel from one seismographic station to another—to gain insight into the earth's structure. Working together, Fehler and Sato planned to use the data to expand their understanding of the waves that cause damage during earthquakes. With new ideas about measurement, they could go back to their respective funding sources for the resources to move their research forward.

The Growth of Global Science

Just as seismologists use noise as an indicator of plate movements that change the structure of the earth, analysts use communications as an indicator of the structure of science. Scientific publications are chief among these indicators. Scientists publish the results of their research to stake a claim to key findings and to enable others to build on their work. This tradition is as old as Robert Boyle's invisible college. The intellectual connections and inspirations that constituted shared knowledge were invisible, but they left traces in the form of the letters, monographs, and pamphlets exchanged among the early college members. Communications leave evidence of actual social and scientific collaborations among researchers; they can be studied to understand science, much as biographers depend on letters to understand the lives of their subjects.[4]

Imagine a map on which a dot represents Fehler as a point in the United States and another dot represents Sato as a point in Japan. Each time Fehler and Sato coauthor a paper, draw an imaginary line, which indicates collaboration, between these two points on the map. Repeat the process for all researchers working in the field of seismology. Now, to see only the communications, take away the geographic map and leave the dots and lines. The result is a network: the dots are nodes, and the lines are links representing communication. This structure illustrates both the opportunities for connections and the constraints that some network members face.

This network can be drawn at a number of levels. The graph we just imagined shows person-to-person links, but once the numbers grow into the thousands, the network can be hard to view. To simplify the graph, we can aggregate all the researchers in the same field and country into a single node and focus solely on the links between researchers working in the same discipline in different countries. Figures 4-1 and 4-2 are drawn at this level, based on articles published in the field of seismology in 1990 and 2000. As in many other fields, international collaboration in seismology greatly increased during that decade. The 2000 network is much more densely populated than the 1990 network. This change is shown by the increase in the number of points and lines and in the greater thickness of the lines, which is proportional to the number of coauthored publications each line represents.[5]

To get a bird's-eye view of global science, we can aggregate even further by combining all of a country's scientists into a single node and studying patterns of collaboration across national boundaries for a particular field. This exercise shows that the share of international scientific coauthorships—that is, publications with authors based in different countries as a proportion of all publications—nearly doubled between 1990 and 2000. In 1990, of all articles published in internationally recognized journals, close to 9 percent were internationally coauthored. By 2000, the percentage of internationally coauthored articles had risen to almost 16 percent.[6] Somewhat remarkably, over the same decade, the number of global papers grew at a faster rate than

Figure 4-1. *Network of Collaborations in Seismology, 1990*[a]

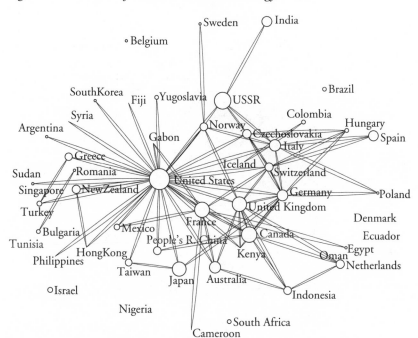

Source: Author's calculations.

a. Country names are presented as they stood in 1990. The isolated nodes had an institution that published an article in the field of seismology in 1990 and was the sole author of the article.

the number of traditional nationally coauthored articles. And between 1980 and 1998, national coauthorships increased by 26 percent, while international coauthorships increased by 45 percent.[7] Other scientists cited these internationally coauthored articles more often than others, suggesting that they might be of higher quality.[8] Similarly, the density of the global science network—that is, the number of ties between nodes in the network divided by the number of possible ties—nearly tripled in the last decade of the twentieth century. As a result of this growth in international collaboration, the average distance between any two scientists in the network is shorter in terms of the number of steps it takes to reach anyone connected to anyone else in the network. On average, researchers collaborating at the global level are between two and four "handshakes" away from each other and benefit extensively from weak ties and small worlds.[9] These connections help bring together complementary or integrative capabilities to create new ideas.

Figure 4-2. *Network of Collaborations in Seismology, 2000*

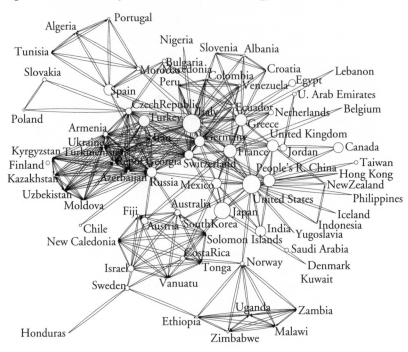

Source: Author's calculations.

Growth across the New Invisible College

To explore the reasons for the shift in the organization of science, I examine four case studies of collaboration in different disciplines spanning ten years. These fields were chosen to test the hypothesis that the organizing factors discussed in chapter 2 (centralized or distributed location and top-down or bottom-up organization) determine whether scientists were more likely to collaborate at the international level.[10] Each of the following fields represents a distinct combination of these factors:

—**Astrophysics** often relies on centralized, large-scale equipment that can be a catalyst for collaboration; it typically gives rise to megascience projects.

—**Mathematical logic** requires no equipment whatsoever, so collaboration arises solely out of the interests of the researchers. Collaboration in this field is typically coordinated (distributed and bottom-up).

Table 4-1. *Summary of Data from Case Studies*

Case study	Number of journals in the cluster, 2000 (base year)	Number of articles published in the journal cluster			Number of internationally coauthored articles in the journal cluster			Percentage of articles in the cluster that are internationally coauthored	
		1990	2000	*Percent increase*	1990	2000	*Percent increase*	1990	2000
Astrophysics	14	4,472	6,547	46	1,301	3,097	138.0	29.0	47.3
Mathematical logic	6	131	309	136	27	117	333.3	21.0	37.9
Soil sciences	10	968	1,382	43	107	453	323.4	11.0	32.8
Virology	9	2,311	2,878	25	327	676	106.7	14.0	23.5

Source: Author's calculations.

—**Soil science** has applications in many countries, but access to certain kinds of soils may require travel and collaboration. As a result, the field is characterized by geotic collaboration.

—**Virology** is a globally distributed research subject like soil science, but it has close links to industry, clinical trials, and disease events. Therefore, collaborative work is both coordinated and distributed, putting this field in the participatory category.

The analysis starts in 1990 with a look at the different levels of international linkages for each of these fields. This is reflected in the number of articles for each field as listed in table 4-1. The four fields show a broad range of levels of activity, both domestic and international. Astrophysics saw the publication of a huge number of global papers. In contrast, there was little international collaboration in mathematical logic. Nonetheless, in each case, the sheer number of papers published at the global level increased by at least 20 percent during the 1990s (see columns I and J). At 136 percent, the increase was greatest for mathematical logic, which started from a very low base. Virology, which was likely already an internationally connected field of science before 1990, saw the lowest increase.[11]

Amazingly, when one starts from the base, participation in international collaboration grows at a very high rate in each of the fields.[12] This suggests that scientists in all fields were interested in creating links beyond national boundaries, regardless of the organizing forces at work. The international links increased in all four fields, whether equipment drove the discipline, whether researchers needed to travel to access resources, or whether they were simply looking for new ideas. In other words—and central to our story—the physical reasons for collaboration were not the determining factor in the growth of global science.

Each of these fields has distinctive characteristics that are likely to influence the decision of scientists to collaborate. Yet all the fields show similar patterns of growth in international coauthorships. Factors like the need to share equipment or to access resources differ across fields. Wolfgang Wilcke studied soil, so he needed to travel a long way to find the specific conditions that would support his research. Haruo Sato wanted to examine the results of an earthquake, so he needed to share data across the globe and to visit certain sites. Luigi Piro was interested in collecting data about the stars from many places around the world, so although collaboration was critical, travel was not involved. Each of these specific circumstances influenced individual decisions about where and how research was practiced in specific fields. But because collaboration grew in all fields, none of these factors can be shown to be the fundamental driver of global collaboration. Whatever factor drove the new invisible college to emerge at the global level, it appears to be one that operated across the board.

Even more striking, if we graph the distribution of international coauthorships within each field, with the number of coauthorships on the horizontal axis and the number of researchers engaged in that level of collaboration on the vertical axis, a power-law distribution is the result (see figure 4-3). Collaboration within each of these fields exhibits the structure of a scale-free network.[13] A few researchers are highly active in international collaboration, but most collaborate only occasionally. The most active collaborators are hubs, and their importance grows over time in response to the dynamic of preferential attachment. In short, all four fields have the underlying structure of a self-organizing complex adaptive system—the same structure that can be found in a forest ecosystem, in the market-based economy, in the human brain, and in many other complex systems.

The Self-Organization of Global Science

Modern science has always had a self-organizing feature, as described in chapter 3. The Royal Society of London organized itself through weak links, small worlds, and the common interests of diverse people. The members of the early Royal Society communicated with each other to exchange ideas, craft methods, and challenge findings. During the next three centuries, communication

Figure 4-3. *Frequency of Publications in Virology, 2000*

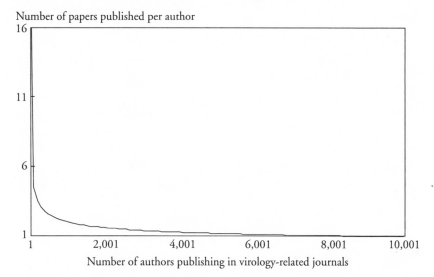

Number of papers published per author

Number of authors publishing in virology-related journals

Source: Institute for Scientific Information and author's calculations.

and collaboration became increasingly important in science.[14] But in the twentieth century, political, geographical, and cultural barriers limited scientists' ability to self-organize into global networks, which made knowledge creation less efficient than it might otherwise have been. During the cold war, for example, the United States and the Soviet Union created virtually duplicate science and technology systems as they tried to beat each other to advances and applications in science and technology.

At a practical level, of course, scientists recognized that serving national prestige helped secure funding for science. But national prestige is not the factor that motivates scientists as they work at their laboratory benches and computers. At the bench scale, scientists seek to solve problems; within social networks, scientists seek recognition for their work and their ideas. These are the forces driving the growth of the new invisible college. To gain insight into the dynamics of science in the twenty-first century, the motivations of individual scientists must be examined.

Scholars such as Robert Merton and Richard Whitely have noted that scientists are fundamentally motivated by a desire for recognition and reward.[15] Scientists advance in their careers by gaining the respect and attention of their peers. Through the reputation attained by publishing the results of their

Table 4-2. *Growth of the New Invisible College, 1990–2005*

Network measure	1990	2000	2005
Number of nodes (countries)	172	192	194
Number of links	1,926	3,537	9,400
Size of core component	37	54	66
Network density[a]	0.131	0.1929	0.2511
Average degree[b]	22.4	36.9	48.7
Average distance[c]	1.95	1.85	1.76
Diameter[d]	3	3	3
Average clustering coefficient[e]	0.78	0.79	0.79

Source: Author's calculations based on data from the Institute for Scientific Information.

a. "Density" is calculated by dividing the total number of links within the network by the number of potential links.

b. "Degree" is the number of links a node has to other nodes.

c. "Distance" is the number of links in the shortest pathway between two nodes.

d. "Diameter" is the maximum number of links required to travel from one node to any other node in the network.

e. The "clustering coefficient" for a node is the number of links between the nodes within its neighborhood divided by the number of links that could possibly exist between them.

research, scientists attract funding, students, and increasing freedom to pursue their work. Ultimately, the independence to pursue one's own research is the holy grail of science.

At the close of the twentieth century, as the cold war ended and the information era began, the abundance of scientific capacity offered unprecedented opportunities for scientists to link to each other. The desire to enhance their reputation and to attract the rewards attached to reputation increasingly pushed scientists to work beyond the walls of their own laboratories and outside their geographic and disciplinary space. In particular, when the Soviet Union fell, eastern European scientists exploded into the global system, creating new links with colleagues in collaborative projects all over the world. As political constraints fell away, the drive for reputation won out over national allegiance, and global science grew at a spectacular rate (see table 4-2). As the table shows, the number of countries represented in a global network of science increased from 172 in 1990 to 194 by 2005, and the number of links more than quadrupled. Yet our analysis has ruled out the Internet, political forces, and the underlying organization of science as the driving forces behind this growth. The remaining explanation is that these changes resulted from the same force that underlies other complex adaptive systems—the operation of a few simple rules.

The Simple Rules for a Complex Network

Recall from the discussion in chapter 2 that simple rules lie at the heart of many highly complex systems and, depending on their initial conditions and resources, can generate an enormous variety of outcomes.[16] Now that we know that the global network self-organizes and we have found that it is a complex adaptive system, the question becomes: Can we identify the simple rules that lead to the complex adaptive system of scientific communications at the global level? This would go a long way toward helping to structure an effective governance system for managing the global network.

The way through the labyrinth is to look for the "if . . . then" rules that lead to the organization of a complex adaptive system. By following this rule, "If the situation shows feature(s) X, then take action(s) Y," an agent within a landscape can adapt and contribute to increased order. In this simple formula, combined with feedback and adaptation, lie the basic elements for creating order out of chaos. The process does not need to be planned in advance; in fact, it is often self-organizing.

John Holland gives an example of such simple rules in his book *Hidden Order*. As he explains: "We see trees over and over again, though we never see even the same tree in the same way. Differing light and differing angles provide a new impression on the eye's retina each time the tree is seen. Still, by dropping details, we see trees in all sorts of contexts and in wonderful variety."[17] In other words, we do not need to try to identify every detail of an unfamiliar species in determining that it is a tree. Instead, we naturally use shortcuts such as, "If the object has a trunk and branches and leaves, then it is a tree."

The primary rules guiding scientific collaboration are also simple. Those seeking new research opportunities reason: "If this connection gives me access to data, funding, or ideas that will advance my research, then I should seek to make the connection." Those approached to provide a resource follow a similar formula: "If this collaboration will help me advance my research or its diffusion, then I should participate in it." These rules operate in all fields of science regardless of their organization or resource base. They explain the formation of the links that make up the new invisible college, as well as the phenomenon of preferential attachment. As a scientist's reputation rises and her access to critical resources such as data, equipment, and funding grows apace, other researchers are increasingly likely to want to form a link with her. The better known she becomes, the choosier she can be when it comes to selecting collaborators. Renowned scientists are widely sought after as collaborators or champions of research. And so top scientists not only have more research partners than second-string researchers, but they also have better collaborators.

Preferential attachment operates across fields as well as within them. Indeed, reputation appears to have an even stronger influence on the formation of cross-disciplinary ties. Because the flow of information is more limited across fields, reputation is one of the few indicators on which a researcher may be able to rely in identifying a trustworthy partner.

This dynamic process of seeking to access resources and build a reputation creates a constant churn of people. Links are made and broken over time. They rarely remain permanent. Because of social obligations that arise in groups, collaborations are more likely to be sustained when researchers work side by side. After all, it is difficult to break off a collaboration when a social obligation exists. This suggests one possible reason why geographically distant relationships—those functioning at the international level—are increasingly attractive and are growing so rapidly. If you work side by side with someone, you come to know what they know and to share a common viewpoint. But if you are seeking new ideas, you need to go outside your usual circle. At the international level, it is relatively easy to break off a relationship that does not prove helpful. In other words, international collaboration may be growing precisely because it is more challenging, yet involves fewer social obligations.

Gaining Entry to the System

The new invisible college self-organizes based on relatively simple rules set and followed at the level of the individual. But this does not mean that the resulting network is simple. Nor does it offer a level playing field. Preferential attachment clearly operates to the advantage of those at the top of the system, whether we think of them as individual scientists or as entire countries. Conversely, it places obstacles in the way of newer entrants with little to offer in the way of resources or reputation. The global science network, then, may be open, but it is not equally accessible to all. Those with weaker connections or weaker reputations can have trouble participating fully in the system. The steep curve of the power-law distribution that characterizes scale-free networks is a metaphor for the difficulty of gaining the attention of those with the most to offer within the network. For example, to gain access to the system, developing countries must leverage the forces that drive networks and learn how to attract the help of scientists around the world.

Brain Drain or Brain Gain?

Early in the morning of August 3, 1997, S. K. Singh, a seismologist at the Institute of Geophysics in Mexico City, received a phone call from a World Bank representative. A devastating earthquake had hit the Jabalpur region of India the month before, and the representative asked Singh if he would join an advisory committee to help India build up its capabilities in seismology research. He quickly agreed. "I hadn't lived in India since I was 20 and I didn't know the region that well," Singh said. "I didn't know if I was so much more qualified than anyone else, but this was a good opportunity to do some interesting science."[18]

With a World Bank grant and funding from the government of India, Singh visited Jabalpur to meet with local seismologists and survey the ground. He also entered into a collaborative research project, inviting two scientists he met in Jabalpur to work with him in Mexico for two months. The relationship in this case was largely one-sided: "Usually collaboration is at the level where each person brings something, but these people didn't really know the technology—even though they were trained scientists," Singh explained. "So in my lab we trained them like students in the use of the latest seismic equipment."[19]

I heard stories similar to that told by Singh from many scientists who are working with colleagues in developing countries. In many cases, the scientists from the scientifically advanced country had a personal as well as a professional reason to work with colleagues from the developing country. These researchers are part of a scientific diaspora that plays an increasingly important role in the new invisible college. Historically, the movement of researchers from developing to developed countries has provoked concern about "brain drain." This term describes the situation when nations with few resources lose their most valuable people—with capable, highly gifted minds—to more developed nations. Educated in more developed countries, scientists and engineers from the developing world contribute to the scientific prowess and economic growth of the countries where they practice their profession, not to those in which they were born. Many analysts have blamed brain drain for the limited ability of many countries to apply scientific and technical solutions to local problems.[20]

The fear of brain drain might have made sense in a world where national governments expected to capture the payoffs generated by the scientific capability located within their borders. But this analysis of the new invisible college suggests that the costs and benefits of the global movement of scientists should be reassessed with fresh eyes. Any country that seeks to tap into the benefits of modern science clearly needs the assistance of skilled scientists, engineers, and technicians. These people need to speak the language of science and understand its norms, in addition to possessing the skills and knowledge associated with individual disciplines. But in a networked world, it is less clear where these people need to *sit* to benefit a particular country or region. Does a country need to have its own scientists to be part of the new invisible college? Do those scientists have to work in laboratories within its borders? Do they need to work in the same place? How do people join a distributed network or team? And how can countries tap, absorb, and apply the knowledge they generate?

FREE AGENTS

Any discussion of the role of individuals in the new invisible college must begin with the observation that each scientist or engineer, each student or postdoc, is the scientific equivalent of a "free agent" in sports. Scientists and engineers are free to follow their own interests and careers wherever those may lead. They cannot be counted on to honor their allegiance to their countries over their allegiance to science and their own careers. Some do, of course, and some will go out of their way to help their country of origin. But policymakers cannot assume such loyalty. Most scientists will seek to enhance their reputations or gain access to resources, regardless of the interests of their nation of origin, and perhaps even at its expense. A country can train its own scientists or engineers, but only with great difficulty can it force them to stay. If they are good, sooner or later a better opportunity is likely to lure them away.

Consider the case of Elena Rohzkova. Growing up in Russia in the 1980s, Rohzkova had the same dreams as many other young women—finding a husband and having a family were high on her list. The difference for young Rohzkova was, as her education continued, she emerged as one of the most brilliant young scientists of her generation. The high quality of her doctoral work in bioorganic chemistry resulted in the offer of a postdoctoral fellowship at Tohoku University in Japan, one of the world's top centers for nanoscience research. Her work in Japan gained her international recognition along with the attention of the Princeton University chemistry department, which offered Rohzkova a chance to spend time there as a special researcher. From there, the University of Chicago attracted her to its research center by offering access to state-of-the-art equipment and collaboration with the medical school—an opportunity that was too good to pass up. At the University of Chicago, she met an American scientist who became her husband. Together they realized her teenage dreams of having a family—only this family was in the United States, where she remains today—still a leading young scientist, still very much a free agent within the new invisible college.[21]

The logic behind this movement was captured succinctly by Anand Pillay, a mathematician trained in the United Kingdom: "In math, people and ideas move, not money! So I moved!" After receiving his doctorate, Pillay could not find his ideal job in England. His search for the right position took him to McGill University in Canada. "If I had decided to stay in England originally, it would have meant doing something else, something different from what I was trained to do," he explained. "By 1986, when the opportunity to return to England presented itself," he said, "there was not much reason for me to return there."[22] Instead, Pillay eventually settled at a university in the United States. In addition, he has had many opportunities to travel and work abroad, including visiting professorships in Japan, Russia, Poland, and France.

One of Pillay's coauthors is the equally peripatetic Saharon Shelah, an Israeli mathematician based at the Hebrew University of Jerusalem and Rutgers University in the United States. Shelah is one of the most active researchers in his field, having published nearly 900 articles with some 200 coauthors. Shelah also has the distinction of having an Erdös number of one, based on his coauthorship of three papers with Paul Erdös, one of the most prolific mathematicians of all time.[23] The Hungarian-born Erdös, who spent much of his career traveling from one research institution to another, wrote roughly 1,500 articles with 511 collaborators.[24]

Erdös and Shelah might be considered additional examples of brain drain, spending much of their careers outside their countries of origin. But their stories also suggest the benefits of circulation. By meeting other researchers and sparking their interest, people join and strengthen their position in the global network of science. Staying home, on the other hand, may mean becoming isolated from new ideas and missing out on opportunities for partnership. Even today, most collaborations begin with face-to-face meetings.[25] Contact can be initiated or continued over the Internet, but such forms of communication alone rarely result in significant collaborative work. The unwritten rules of group participation require that people physically work together, at least for some period of time. Once this contact has been made, people can use virtual connections to continue collaborative research.

In such collaborative relationships, each member of the team is typically self-funded. In other words, the researcher is funded, not the project. As Anand Pillay points out, people and ideas move more readily than money.

Because governments supply money for science, it is logistically difficult to move money from, for example, country A to a scientist in country B. But people and ideas circulate easily around the globe, aided by the ease of traveling by air and the even greater ease of communicating via the Internet.

Such movement and interaction not only drive the growth of the network, but they also foster the growth of individual careers. This is particularly true for graduate students, who function as the knowledge equivalent of "messenger RNA," according to one scientist.[26] Many of the researchers I interviewed told me that one way they stay on top of research is by sending their graduate students to research sites around the world. Students collect data and sit up all night with their colleagues watching experiments. They form valuable connections and carry vital information from institution to institution. Once they are established in their careers, they can help to seed scientific capacity in their home countries.

GOING HOME

Increasingly, students from developing countries are choosing to return home once their studies are complete. According to the U.S. National Science Foundation (NSF), in 1980, 47 percent of Chinese students earning a Ph.D. in the United States reported that they had firm plans to stay in the United States. By 1993, the total number of Chinese doctoral students in the United States had increased, but the percentage planning to stay in the country had dropped to 45 percent.[27] Similarly, the percentage of students from India who planned to stay in the United States dropped from 59 percent in 1980 to 50 percent in 1993. South Korea's numbers were even more dramatic: in 1993, only 18 percent of those studying in the United States planned to stay in the country after receiving their degrees, compared to the 41 percent of students who planned to stay in 1980. In all these cases, economic development has created new opportunities at home for talented researchers trained abroad.

In addition, as Singh's example shows, many expatriate scientists find ways to contribute to scientific development in their countries of origin. The RAND Corporation conducted a survey of 100 U.S.-based scientists, finding that as many as one-third of the scientists who were collaborating internationally were doing so with someone in their country of origin.[28] These foreign-born scientists and engineers were also more likely to accept and train talented people from their home country, fueling the cycle of knowledge creation and capacity building. This long-term cycle takes years to bring to fruition, but it is clearly working to the advantage of many countries, such as

Vietnam, China, Mexico, and South Korea, that have made investments in basic capacity.

Foreign-born researchers serve as important human bridges, catalysts, and financial links between the developed and developing worlds. By conducting collaborative research with counterparts in their country of origin or by serving as science advisers to private or public organizations, these foreign-born scientists are helping to advance scientific capacity in developing countries, often with funds from scientifically advanced countries—a phenomenon that can justly be labeled "brain gain."

AnnaLee Saxenian makes a similar argument in her work on Silicon Valley. Countering fears of brain drain, she finds that many of Silicon Valley's foreign-born researchers ultimately return to or form partnerships with companies or other organizations in their countries of origin. These circulating scientists bring valuable know-how, experience, and contacts back to the developing countries where they were born.[29] In the process, they contribute to the creation of transnational technical communities that facilitate the spread of knowledge around the globe.

In science and technology, then, the question facing a developing nation like Vietnam or India is not how to keep smart people home. It is not even how to get them to *return* home. The real challenge is how to get a country's researchers into the new invisible college and then attract other researchers to work on local problems. Many governments have begun to address this challenge by sending students abroad for advanced training. According to the NSF, the number of science and engineering doctorates awarded in the United States to non-U.S. citizens grew from 5,100 in 1985 to 9,600 in 2001.[30] During the same period, foreign students in the United States earned close to 148,000 doctorates in scientific fields, and the share of such degrees awarded to foreign nationals rose from 26 percent in 1985 to 35 percent in 2001.

More broadly, the United Nations Educational, Scientific and Cultural Organization (UNESCO) estimates that the United States hosted 583,000 foreign students in 2001 and 2002, making it the leading recipient of students from other countries.[31] Around 30 percent of all foreign students study in the United States, and roughly half study in Europe. Together, the United States, the United Kingdom, and Germany host half of the world's foreign students. Add the next two leading host countries (France and Australia) and these five countries serve two-thirds of the world's foreign students. In contrast, few students travel to less developed regions. South America is the least common destination for foreign students (hosting only 0.4 percent of the global foreign student population), followed by Africa (1.2 percent).

Asians account for a high proportion of those studying abroad, particularly at an advanced level. More than 60 percent of foreign students in the United States come from Asia, as do four of every ten students pursuing work at the tertiary level in a foreign country. (Three of every ten are European, and one of ten is African.) Students from developed countries, on the other hand, are more likely to stay close to home. North Americans make up less than two percent of the total number of foreign students, and eight of ten European students study in another European country.

Joining the Network

According to the Institute of International Education, about 45 percent of foreign students are in a technical field, suggesting that over the longer term, many will be able to contribute to the development of local scientific capacity.[32] Developing countries and nongovernmental organizations should promote this process by encouraging and funding bright young students to study abroad. But science policy cannot stop there. Ideally, scientific organizations should also keep track of where people go and stay—a task that is greatly complicated by privacy concerns. In addition, to ensure that circulation produces brain gain and not brain drain, governments need to make collaboration with domestic researchers attractive both to the free agents who go abroad and to the new invisible college at large. This might involve defining interesting research challenges, making unique data or resources available, offering funding for targeted projects, or hosting conferences that create opportunities for leading scientists to focus on local opportunities and problems.

All these policies have a common goal: joining the network. The network is the critical resource. The best way to get knowledge to the people who need it is by broadening the scope and reach of the new invisible college. Instead of trying to control the circulation of researchers and ideas, policymakers need to focus on building an environment that encourages researchers to self-organize in pursuit of answers to important problems. Allowing researchers to find the places where they can do their best work and encouraging them to pick and choose collaborative opportunities will increase the efficiency of the knowledge system as a whole. But the problem of then funneling that knowledge to the places where it is most needed will remain. Solving this problem requires a better understanding of the geography of knowledge, which is the subject of the next chapter.

CHAPTER FIVE

THE VIRTUAL GEOGRAPHY
OF KNOWLEDGE

> We enter a new territory. It would be presumptuous to suppose that we
> would understand a new continent when first alighting on its nearest shores.
> We are seeking a new conceptual framework that does not yet exist.
> Nowhere in science have we an adequate way to state and study the inter-
> leaving of self-organization, selection, chance, and design. We have no ade-
> quate framework for the place of law in a historical science and the place of
> history in a lawful science.
>
> STUART KAUFFMAN, *At Home in the Universe*

The previous chapter emphasized the role of free agents, those individual
researchers who can move freely within the global network, in the
invisible college. The freedom to pursue interests and opportunities around
the world has done much to break down the national focus of science that
dominated for much of the twentieth century. Geography does, however,
retain an important, if changing, role. The invisible college does not exist in
any one place—by definition it encircles the globe. Even so, science is practiced more intensively in some places and less so in others.

Some projects are best carried out in particular locations. Others could
potentially be located anywhere but might require the co-location of many
individuals and activities, at least if research is to be conducted on an efficient scale. And even when distributed collaboration is a viable option, face-
to-face meetings remain a critical starting point or an important part of
many projects. Despite the growing efficiency of communications technology, it cannot replicate all the benefits offered by working side by side. This is
true both at the level of individual researchers—who absorb valuable (if

Stuart Kauffman, *At Home in the Universe: The Search for the Laws of Self-Organization and
Complexity* (Oxford University Press, 1995), p. 185.

immeasurable) tacit knowledge by working in the same lab—and at the level of economies, which benefit from the spillovers generated by geographic research clusters.[1]

In this chapter, then, the focus shifts from people to place. The chapter reviews the current geography of science and the factors that drive the distribution of scientific activity and explores why this distribution matters. It also examines how the benefits of science can be distributed more fairly. The discussion begins with a look at how place influences collaboration at the individual level.

From Brazil and China to the United States

During the 1990s, Frank E. Karasz, who is now the Silvio O. Conte Distinguished Professor at the University of Massachusetts in Amherst, generated a great deal of excitement in the field of polymer science by engineering a material he called "conjugated polymers." By passing an electrical current through this new type of polymer, he could cause it to emit light. This would be like lighting a dinner plate—an object that does not conduct electricity—with a battery and a wire. Karasz's discovery revolutionized the field of polymers and optics and offered the possibility of developing new forms of lighting and lightweight displays. It also won him numerous awards and citations, which placed him at the top of his field. His appeal as a collaborator soared.

In 2000, Karasz published a frequently cited paper on the characteristics of light-emitting conjugated polymers.[2] The coauthors of the paper hailed from several countries. D. E. Akcelrud was a Brazilian scientist on a yearlong sabbatical in Massachusetts. She had met Dr. Karasz when he visited Brazil on a scientific exchange. Akcelrud brought with her M. R. Pinto, a postdoctoral researcher from the University of Rio de Janeiro; his contribution to the 2000 paper was to synthesize the polymers for the experiment. In addition, a Chinese postdoctoral student, Bin Hu, contributed to the project by analyzing the results.

Akcelrud subsequently returned to her university position in Brazil, where she continues the line of research she began in Karasz's laboratory. She mails samples of synthesized polymers to the Massachusetts laboratory for measurement and receives the results via the Internet. She also keeps in touch with Bin Hu, who returned to China to conduct materials research. Together they have created a cluster of polymer research within the new invisible college.

Karasz is an example of a highly attractive node in the global network of science. Akcelrud connected with Karasz to improve the quality of her own research, and continues to work with him, even though they are no longer working in the same laboratory. Akcelrud, in turn, was the link between the young Pinto and Karasz. On his own, Pinto would not have been able to connect with Karasz, but thanks to the small-world phenomenon, he gained access to and the opportunity to work with one of the top scientists in his field.

The experience of this research team is replicated in thousands of connections among scientists seeking complementary capabilities, reputation, and resources. Researchers gain knowledge and capabilities by tapping into this system through strong and weak links. The results of their research are readily shared through a system of publications, and in theory at least, scientists can access these findings from anywhere in the world when and where they might be useful. The theory breaks down somewhat, though, because many journals are available only through a subscription service, which may put them out of the reach of researchers in universities that do not have large enough budgets to allow access to the full range of scientific literature.

Yet to cement their connection, the individuals who made up this team had to come together in a small college town in western Massachusetts, where Karasz, a Slovak by birth, had established his laboratory. Not only did Karasz have the expertise that drew them there, but he also had the specialized equipment. "The University of Massachusetts has the best polymer science section in the country," he says proudly. "Our laboratory has state-of-the-art equipment for advanced polymer research."[3] In many fields, the availability of such equipment, as well as the proximity of top researchers, has an important influence on where collaboration takes place.

The Global Distribution of Science

Such factors, combined with the forces of historical accident that bring scholars the caliber of Karasz to pursue their livelihoods in the United States and the greatly uneven global distribution of power and wealth, have led to the concentration of scientific activity in a relatively small number of countries, as shown in figure 5-1. In this map, the height of each line is determined by the number of papers that were published by researchers based at local institutions. The good news is that scientific talent appears in many places in the world. The potentially bad news is that it is highly clustered in the developed world. In this respect, the world is far from flat. The peaks on the map around such cities as Boston, London, and Tokyo represent the results of significant, long-term national investment in science. Such investments have

Figure 5-1. *Scientific Activity Is Concentrated in Relatively Few Countries*

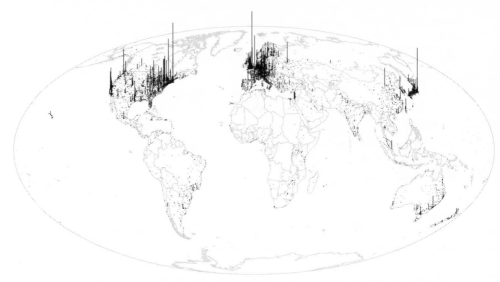

Source: Artwork by W. Bradford Paley; data provided by Richard Klavans and Kevin Boyack.

long been possible only for rich nations. In 2004, just fifteen countries spent 90 percent of the world's research and development (R&D) dollars.[4] Before 1960, six countries accounted for the same proportion of the world's R&D.

Within these countries, the geographic distribution of scientific spending and capacity is also uneven and highly concentrated. In 2003, for example, almost two-thirds of R&D spending in the United States took place in ten states. California alone accounted for more than one-fifth of the country's $278-billion R&D spending in that year. Also in 2003, more than half of all privately funded R&D by computer and electronic products manufacturers took place in just three states—California, Massachusetts, and Texas.[5]

As these examples suggest, each field within science has a different geography, and the extent to which activity is concentrated in a particular place varies by discipline. For example, fields that require complex, expensive equipment—those fields with very strong "gravitational pull"—will cluster around a few research centers. Nonetheless, research across most fields is generally located in or near big cities in wealthy countries. The same holds for the institutions of higher education that train students and conduct world-class research. Within scientifically advanced countries, these academic institutions are highly concentrated in a few regions. The resources to create knowledge, in short, are not equally spread.

What's more, much of the knowledge created in these clusters is not easily accessible to others. Published articles are often in limited-access journals with subscription fees that are out of reach for institutions in poorer countries. Intellectual property laws protect some information. National governments also limit access to science, as noted earlier, because they view knowledge as a national asset to be kept close to home.

Explaining the Geography of Science

Numerous factors contribute to the uneven distribution of scientific activity, including the history of government investment in the wealthier countries. Obviously, richer governments can provide greater support for science, and understandably, they prefer to spend their research funds within their own borders. Moreover, in choosing exactly where to allocate facilities, policymakers often consider political goals. In determining where science takes place, however, nonpolitical factors also come into play. These include the capital requirements of a particular research project; the extent to which it depends on access to unique resources; historical accidents; and such phenomena as preferential attraction, economies of scale and scope, and stabilization into a paradigm (known as "lock-in").

Some fields require expensive equipment or necessitate that research be conducted at particular locations. In such cases, organization around a few centers of excellence is natural and efficient. The knowledge system does not need and cannot afford a synchrotron in every country; nor does it make sense to distribute centers for research in oceanography randomly or evenly around the world.

In some cases, such as oceanography, the location of concentrated investments is shaped largely by geography. But in others, it may be determined by historical accidents or investments that make a particular city or region a leading candidate for additional investment. For example, Hannover, Germany, has a centuries-old reputation as a center for chemical research. Geneva, Switzerland, has long been a center of physics research, a history that made it a natural choice for the location of the European Organization for Nuclear Research (CERN), the world's largest particle physics laboratory. Early on in the development of twentieth-century astrophysics, Santiago, Chile, became the leading site for telescopes trained on the southern sky, and local research centers have capitalized on the equipment to become leaders in astronomy. Because of its long history of studying earthquakes, Kobe, Japan, invested in a shake table along with other facilities for seismology research that draw scientists from around the world.

As these examples show, once significant investment has been sunk into a particular region and cumulative advantage adds to the value of the initial investment, the region takes on a certain gravitational pull within the network. Consequently, the ultimate distribution of scientific activity is "path-dependent."[6] Several related forces help account for this effect. Because of economies of scale and scope, it is often more efficient for different research projects to be conducted in the same location. In turn, these economies of scale and scope can be traced to both tangible and intangible factors, such as expensive equipment and the accumulation of social capital and tacit knowledge. A preexisting stock of social capital makes it easier for investigators to form research partnerships, and tacit or embedded knowledge helps make those partnerships more productive. At the level of an institution or research cluster, they also help lock in an early head start.

Preferential attachment on a personal level reinforces cumulative advantage at the field level. Researchers like D. E. Akcelrud naturally prefer to work with those who are at the top of their field like Frank Karasz. As a result, centers of excellence benefit from the constant inflow of highly productive and well-funded scientists. In this way, they successfully balance two forces that any successful institution must accommodate—stability, through the agglomeration of knowledge, and variation, through the arrival of new participants and new ideas.

Why Does the Location of Science Matter?

World-class scientific facilities have long conferred prestige or bragging rights on the countries that support them. If that were the only benefit they offered, the geographic distribution of scientific activity might not be of great concern. But concentrated scientific investment also yields more significant payoffs. At the national level, scientific investments are clearly correlated with economic prosperity, as many have pointed out in studies on both science and economic growth.[7] Yet a causal connection between science and prosperity has never been convincingly made in either economics or science studies. Science may be a catalyst for development that kicks in once some level of economic growth has been achieved.

Within nations, clusters of scientific activity can generate significant economic spillovers, not only by creating employment opportunities at and around scientific institutions, but by creating knowledge that can be captured and exploited by local entrepreneurs. In many places—notably in Silicon Valley—research institutions clearly serve as wellsprings of knowledge in a way that creates connections to regional growth. Much has been written about the regional spillover of research into local technology development and the larger impact of this spillover on regional economic development.[8] Together, research institutions, businesses, and the local population appear to create a virtuous cycle that promotes a fertile environment for both scientific and economic activity. This finding has led government actors at all levels (municipal, state, national, and international) to promote the development of local clusters with the goal of setting such virtuous cycles in motion.

Such clusters can do more than foster economic growth; they can also deliver scientific and policy benefits in the form of knowledge that is targeted to local problems. The needs of countries that are home to leading scientific institutions are more likely to be addressed. On the other hand, countries that cannot afford to make large-scale investments in research often find it difficult to interest the outside world in their concerns. This dynamic helps explain, for example, why vast resources are devoted to combating cancer in the United States, while diseases like malaria, which afflict far greater numbers of people in developing countries, have traditionally received little attention. The concentration of scientific activity means not only that some areas and some countries generate more knowledge than others, but also that those same areas and countries get to determine what kind of knowledge is created and what kinds of problems are solved.

This is true not only of research institutions with global reputations; it also applies on a smaller scale. Most scientifically advanced countries support research laboratories dedicated to agriculture, health (including food safety), the environment, biotechnology, manufacturing, and transportation that focus on exploring and solving localized problems. Such laboratories play a critical role in supporting the quality of life in these countries. But unfortunately, the knowledge they hold is often so localized that it cannot be effectively applied in other settings.

Redistributing the Benefits of Science

Given that scientific activity offers such significant benefits, two questions arise. First, to what extent can virtual connectivity make up the gap between the places where scientific resources exist and the places that need those resources? Second, what else can and should be done to distribute those resources or their benefits more equitably?

The Role of Virtual Connectivity

The growth of the Internet and the evolution of the World Wide Web have clearly increased the ease with which researchers can share data and resources that enhance creativity. Today scientists and engineers can find each other and communicate more readily than ever before, and they can use the Internet and the Web to share, store, and improve on data. Yet despite their impressive magnitude, these changes have not revolutionized the way in which science is done. In interviews with scientists, it quickly becomes clear that even though the Internet and the Web have greatly increased the efficiency with which scientists could share information, for the most part, they did not provide a *new* capability. Researchers shifted from letters to e-mail. Instead of storing data on paper and analyzing the information by hand, they now store it in digital forms and conduct computer-based analysis.[9] Once collecting databases might have taken months. Now, these databases can be compiled and augmented in a matter of hours.

There have been some exceptions to this rule. For example, distributed or grid computing has it made possible for projects that require enormous computing power to farm out work to thousands of ordinary personal computers instead of relying on access to supercomputers. The Search for Extraterrestrial Intelligence (SETI), a project based in Berkeley, California, uses this approach to examine the vast amounts of data collected by the Arecibo radio telescope in Puerto Rico. The observations collected by the telescope are mailed to SETI's facility in Berkeley and then broken into small chunks, which are distributed across the Internet to volunteers who have downloaded SETI@home software. The software automatically analyzes the data and sends the results back to Berkeley, where they are integrated into a huge database.[10]

Other efforts have been made to tie together research laboratories that use large-scale equipment to create a virtual network. Such networks have been created for nanoscience research, for example. Because the nanosciences study materials at the atomic and molecular level, researchers require highly specialized equipment—typically advanced microscopy—in order to see the subject of their research. Currently, work is advancing to use atomic force microscopy nanoprobes for teleoperated physical interactions and manipulations. For example, the nanoManipulator, a robotic system developed at the University of North Carolina at Chapel Hill, allows three-dimensional visualization of real-time experiments over the Internet.[11]

In theory, these types of tools should make resources accessible to distant researchers who cannot travel to the institution where the tools are based. This has often been offered as a way to use the Internet to link poorer countries to knowledge centers.[12] But this strategy is unlikely to succeed on its own. Virtual links work best when researchers have already worked face to face and then use computer and telecommunications tools to cooperate over communication networks, as exemplified by the conjugated polymers team.[13]

Moreover, being able to access equipment is not the only problem. An equally important challenge is finding a way to access the dynamic discussion and interaction—the social capital—that comes with working in a laboratory with other researchers. Researchers can tap into such discussions through electronic communications. But these mechanisms rarely capture the full range of information that is exchanged in casual, freewheeling face-to-face discussions, especially those involving many people. In addition, they cannot convey tacit knowledge—the intuitive experience that researchers may have and share without even being aware that they have done so. By working side by side and learning by doing, scientists pick up shortcuts, habits, and best practices that advance their work, possibly without ever realizing what they are doing. If these practices are so instinctual or ingrained, the researchers might never think to put what they are learning into words for wide dissemination.

This oversight applies to more than research techniques. Knowledge creation also requires learning the norms and values of the dominant research community, as well as the language that community shares. These norms are rarely codified either electronically or in print. Instead, they are transmitted through practice, example, and casual comments—all methods of communication that virtual connectivity cannot access.

Remapping Science: A Thought Experiment

Given the continuing importance of where science happens, what can or should policymakers do to try to distribute scientific resources more equitably? To address this question, let's undertake a thought experiment inspired by John Rawls's influential theory of justice. In his 1971 book of the same name, Rawls suggests that to arrive at a fair ordering of society (or, in this case, a fair distribution of resources), we should deliberate behind a "veil of ignorance" that prevents us from knowing the details of our own status. Behind the veil, everyone has an equal chance of being rich or poor, clever or backward, feeble or strong. Rawls describes this starting point as the "original position," or "the appropriate initial status quo which insures that the fundamental agreements reached in it are fair."[14] He argues that if we start in this original position, we will end up with a general conception of justice as fairness, which "requires that all primary goods be distributed equally unless an unequal distribution would be to everyone's advantage."[15] By agreeing to this rule, each individual in the original position can maximize his expected well-being in society.

Applying this proposition to science is a tricky matter. Recall from the previous discussion that the concentration of resources often contributes in an essential way to knowledge creation. As a result, an unequal distribution of activity would seem to be good for science and, at least theoretically, to everyone's advantage. At the same time, however, we are likely to agree that concentration should not be taken to an extreme. No single political entity should become so far advanced over other nations that it draws the vast majority of available resources to itself in a kind of cumulative-advantage-gone-wild scenario, even if the result of the concentration is a great deal of new knowledge. Such a monopoly would mean that other places would lose access to learning opportunities and innovative capability.

The challenge, then, becomes establishing the right balance between the goals of equity (which favor distribution) and those of knowledge creation (which in many cases favor concentration). This challenge is especially complicated because the socially optimal degree of concentration is likely to vary from field to field. Clearly there are cases in science where shared equipment works well. But in other cases, research capabilities simply must be locally available if research is to advance. The same principle of distribution cannot be applied across all disciplines.

To address this challenge, let's return to the four-part schema outlined in chapter 2. There, by juxtaposing the way in which research is organized (top-down or bottom-up) against the way in which it is conducted (centralized or distributed), four categories of scientific projects were derived: megascience, geotic, coordinated, and participatory. Recall that megascience and some geotic projects tend to be very "heavy" and require large-scale, specialized equipment. These projects are very expensive and become essentially sunk in place once created. It would be highly inefficient for the knowledge system to invest in more than a few of such resources (or sometimes even more than one). In such cases, we can conclude that a highly uneven distribution is acceptable, although fairness would be better served if these centers were open to users based on a sliding user fee and through virtual links. In other words, access to these resources should be open, and the knowledge developed within them should be widely shared.

In contrast, coordinated and participatory activities have a lower cost of entry. The smaller scale of such research means that many laboratories can be located around the world. In this case, a variety of distributions may be socially optimal. For example, similar laboratories could be spread relatively evenly across the landscape based on population, enabling easy access for most researchers. Or laboratories in different regions could specialize in particular areas within the same discipline and participate actively in the exchange of resources and information based on their capacities and needs. This second approach makes sense when local conditions are an important research input. For example, in agricultural research, small, local laboratories and extension centers are often the best mode for conducting research and transferring knowledge.

The challenge of ensuring experiential learning in a distributed environment, however, must still be addressed. A large component of knowledge creation involves tacit learning. If research is compartmentalized to enable shared tasking, the issue becomes one of discovering where the knowledge is integrated and exploited. If knowledge integration takes place in a single favored location, important parts of the learning process may be lost to other members of a distributed team unless they find ways not only to access the integrated knowledge, but also to "tie it down" to meet specific needs at the local level. As an example, the research conducted for the Human Genome Project was distributed and coordinated around the globe. The knowledge was shared with all participants in the research. But the ability to integrate the knowledge into findings and to turn findings into products was limited to those parts of the world that could integrate these functions. Only a few places met these criteria.

In summary, the uneven distribution of science is not the problem in itself—even in a just system, some fields of science would be unevenly distributed because of the scale and scope of the investments required to advance knowledge. In these cases, knowledge creation in a specific field requires that the contributing factors be geographically concentrated. The problem lies in the inability of certain places to integrate knowledge and direct it toward solving problems. Although governments often supply these functions as public goods, inequalities arise because of the biased structure inherent in the national systems that inhibit knowledge diffusion, not because of the location of the research itself. This can be partly remediated through open access and virtual networking and partly by changes in government policy.

The greatest inequity in the location of research applies generally where concentration delivers limited benefits and particularly where knowledge must be locally available and relevant if it is to be effective and useful. In these sciences—such as soil science, agriculture, aquaculture, biology, and hydrology, among others—efforts must focus on ensuring that research capabilities are made available, ideally in the form of targeted local investments that are linked to important nodes within the network of science. Such investments can critically boost a region or country's ability to absorb and build knowledge.

Back to the Real World

What can this thought experiment teach us about the policies and priorities that should guide science in the twenty-first century? First and most simply, wherever possible, artificial barriers to the transfer of knowledge should be removed so that when the opportunity arises, scientists can form connections and knowledge can flow and grow. Such barriers include the costs associated with accessing scientific journals, attending conferences, buying equipment, and traveling to location-specific research sites. In addition, knowledge transfer should be actively promoted through increased use of information technology and the application of funds to help spread know-how to developing countries. It is equally important, though, to acknowledge the limitations of this strategy. Some knowledge can be easily transferred and some cannot.

Although some knowledge can be easily absorbed locally, most fields require extensive training before they can be understood and even more training before they can be applied to development challenges. More important, science is not simply a body of knowledge that can be transferred from one place to another to solve problems and meet challenges. Fundamental science, or scientific research, is the process of contributing something truly new to the knowledge system. This process is inherently unequal and "exclusionary" in that the self-organizing forces of the new invisible college inevitably favor some connections and patterns of exchange over others.

How then can most of the world's countries participate more fully in this system? In the past the answer to this question has focused on creating home-grown scientific capacity and in particular, developing centers of excellence that emulate those of the developed world. In the future, the solution will lie in developing more nuanced strategies to both sink local investment and link to existing resources. These strategies will build on and reshape the physical geography of science as well as the virtual geography of knowledge.

Toward a New Geography of Knowledge

For much of the twentieth century, policymakers operated on the assumption that to participate effectively in science at the global level, a country needed to create the institutions, the range of sciences, and the skilled workforce that support the knowledge systems of the scientifically advanced countries. Many development initiatives focused on aligning institutional and legal organizations to create national innovation systems similar to those found in advanced countries. Not surprisingly, efforts focused on building national innovation systems have been complicated by the fact that different countries have pursued different paths toward scientific and technological development. In Japan, for example, most research capability emerged from within the private sector. Within the United Kingdom and the United States, however, research capabilities were largely developed with government support. In Europe, governments determined standards for materials being used in research and manufacturing; in North America, the private sector took on this role.

Other scholars have sought even more fundamental explanations of the degree to which different economies have succeeded in creating and using knowledge. For example, Francis Fukuyama suggests that the level of associational trust within a society has an important influence on the level of cooperation, which is in turn needed for economic growth and technological change.[16] This is an intuitively appealing idea because, as noted earlier, trust and social capital are critical to the new invisible college's ability to create knowledge through integrative teams that reach across traditional political, disciplinary, and geographic boundaries. In practical terms, however, fostering associational trust is an even more daunting task than replicating another country's basic legal, political, and economic institutions.

A slightly less ambitious approach might be to replicate major elements of a wealthy country's scientific establishment. Again, this strategy has some surface appeal. Many countries have promoted their own prestige projects in the hope of developing or attracting top researchers and capturing the economic and scientific benefits of a research cluster. But many such efforts failed because they neglected to identify unique local conditions that could draw top scientists and support the generation of new knowledge. A newcomer to the system cannot hope to rival the United States or Europe in high-cost disciplines, such as particle physics, in which the world's wealthiest countries have already made extensive investments. But by focusing on a unique local resource or a uniquely compelling problem, the newcomer may be able to build a new hub within the new invisible college.

Efforts to create effective institutions and local capabilities remain an important part of science policy in the twenty-first century. But such efforts must be placed in a new context. Instead of thinking of their scientific investments as existing in isolation, policymakers need to visualize them within the context of the virtual geography of knowledge. In other words, after scanning the structure of the new invisible college and identifying local needs and opportunities, policymakers must determine when they should link to existing resources that are available but geographically distant and when they should sink local investment and build or enhance local or regional capabilities. In many ways, this choice is similar to what business strategists call the "build or buy" decision: When should a firm build new capabilities or keep certain functions in house, and when should it rely on the market to provide important inputs or services? In both cases, decisionmakers must weigh a broad range of factors, including the scale and scope of the required investment, the degree to which their needs are distinctive or even unique, and their existing capabilities.

A key difference between the two situations, however, is that when they consider investing in science, policymakers need to see their choices as "link *and* sink," not "link *or* sink." These two approaches are largely complementary. It is often necessary to sink some local investment to create the ability to link to the global system (by becoming an attractive partner) or to ensure that those links pay off. Within the global knowledge network, the challenge is to tap into and tie down knowledge locally. Often that process is effective only if some scientific capacity exists on the ground, in the form of institutions and researchers that can appropriate, apply, and build on knowledge developed elsewhere.

This is not to suggest that policymakers should seek to construct science according to a plan—quite the opposite. Constructing scientific teams to meet political goals introduces inefficiencies into the system. The built-in rules and emergent structures of the new invisible college have proven to be highly effective in making the connections needed to generate knowledge. Consequently, planners should seek to create and locate incentives to encourage the kind of organization and investment that will prove to be locally sustainable.

Sinking and Linking in Uganda

Forward-looking policymakers, such as Peter Ndemere, the executive secretary of the Uganda National Council for Science and Technology (UNCST), are rising to this challenge. Ndemere worked for years to formulate Uganda's science and technology plan, to collect data to support it, to build consensus around the document (which was issued in 2007), and to budget for an increase in science and technology. Throughout this process, he showed a keen awareness of the need for Uganda to develop a strategy that not only is highly tailored to the country's circumstances, but also draws on international connections. As Ndemere acknowledged, "Uganda will need to tap into a lot of knowledge we don't have within our borders."[17]

As an important first step, Ndemere and his colleagues helped to guide policymakers in selecting priority investment areas. As one of the poorest countries in the world, Uganda cannot support all the fields of science that are found in a scientifically advanced country. It will have to choose its bets carefully. Policymakers cannot afford to invest across the board and see which projects succeed and which fail. Conscious of these constraints and based on substantial public input, Uganda's government has made its initial investments in coordinated and distributed science activities in virology and biotechnology.[18] These areas were chosen because they address local needs and because they offer opportunities for both local and international research. "If we in Uganda want to move beyond just agricultural research, we need to make some bold investments," Ndemere explained, "but we must find the right mix of subjects that will appeal to our own scientists. Otherwise the plan won't go forward. And we cannot go it alone. We have to link up with other groups outside Uganda. Just how to do this? Well, that part is still unfolding."[19]

In addition, if their strategies are to be sustainable, policymakers must ensure that a basic support system for scientific activity is in place. In this respect, they may have an advantage over earlier entrants to the new invisible college. Uganda might not need to invest in the full range of institutions, support services, and functions that scientifically advanced countries have traditionally provided. To some degree—particularly in coordinated and participatory sciences—networks can substitute for homegrown institutions. Computer-mediated communications can help networks fill this role. Moreover, intergovernmental or nongovernmental organizations can provide some of the essential functions that underpin scientific capacity at the regional level, instead of within each country. Understanding these functions and the capacity they support is the goal of the next chapter.

CHAPTER SIX

SCIENTIFIC CAPACITY
AND INFRASTRUCTURE

> This disparity between the rich and the poor has been noticed. It has been
> noticed, most acutely and not unnaturally, by the poor. Just because they
> have noticed it, it won't last long. Whatever else in the world we know
> survives to the year 2000, that won't. Once the trick of getting rich is
> known, as it now is, the world can't survive half rich and half poor.
>
> C. P. SNOW, *The Two Cultures and the Scientific Revolution*

C. P. Snow's famous essay, *The Two Cultures and the Scientific Revolution,*
analyzed the problems created by the lack of communication between
scientists and nonscientists in the early years after World War II. In large
part, the essay realistically assessed the tensions between science and society.
Yet Snow was utterly wrong about the power of science to close the gap
between rich and poor.[1] The year 2000 has come and gone and most analysts
think the gap has expanded rather than narrowed. In addition, new gaps,
such as the digital divide, are emerging.

Instead of diminishing these inequalities, science may have contributed to
their growth. As economist Jeffrey Sachs points out, the countries that did
relatively well in the second half of the twentieth century had higher food
productivity and literacy rates, and lower infant mortality and total fertility
rates.[2] Scientific advances formed the foundation for each of these positive
developments, but under the influence of scientific nationalism, access to
these advances was largely limited to a fortunate few.

C. P. Snow, *The Two Cultures and the Scientific Revolution* (Cambridge University Press,
1959), p. 44.

Today the new invisible college transcends national borders. It connects researchers from the four corners of the world and allows them to combine and recombine into teams that put scientific interests above national allegiance in deciding where and how to work. Scientific nationalism is steadily eroding as self-organizing networks take on the organizing and coordinating roles that national science ministries used to play. Yet the nation-state is far from dead; it remains a key part of the scientific landscape in the twenty-first century. People still live within countries delimited by geographic boundaries and if they are lucky, they live under the rule of sovereign governments. The ability of ordinary citizens to share in the benefits of science is largely shaped by their country's scientific capacity. This ability is determined in turn by the availability of access to the essential functions and services needed to support scientific capacity. Taking these factors into account, this chapter focuses on two questions: What is scientific capacity, and what type of infrastructure is necessary to support it?

Scientific Capacity: Climbing the Ladder

Scientific capacity involves absorbing, applying, creating, and retaining knowledge about the natural world. Generally speaking, these tasks represent a ladder of complexity. For example, it is possible to absorb knowledge without being able to apply it because specialized skills are lacking. When hemorrhagic fever strikes a region of Africa for the first time, scientists may know about the virus but be unable to deal with its consequences. Specially trained technicians may be needed to treat people and isolate a virulent virus, particularly if cultural issues are involved in treatment. In this case, simply having knowledge about treatment may not be enough.

To overcome this problem, policymakers often invest in specialized knowledge transfer services. For example, the agricultural research service providers in Uganda work under the umbrella of the country's National Agricultural Research Organisation (NARO) to address specific, local problems. A local producer may come to NARO to conduct collaborative research with NARO scientists as well as neighboring countries and international researchers to aim at goals such as improving local seed yields. To ensure that knowledge can be applied locally, NARO works with the Ugandan National Agricultural Advisory Service (NAAdS) to train and instruct rural farmers in planting techniques resulting from the research at NARO.[3]

It is also possible to absorb and apply scientific knowledge without creating it. If an oil tanker wrecks off the coast of Canada, cleaning up the oil spill requires scientific knowledge about the properties of oil, marine environments, and regional wildlife. This knowledge can be absorbed and applied locally in response to the urgent situation without creating new knowledge or retaining knowledge for future applications. Once the oil spill is cleaned up, the community may simply hope that it will not need that kind of specialized knowledge again.

Knowledge creation is a still more complex aspect of scientific capacity, requiring the ability to define the questions and the experimental approaches needed to push beyond existing knowledge. The level of knowledge needed to conduct creative experimentation grows more detailed and intricate as a field of science evolves. This is especially true in areas involving cross-disciplinary fertilization. According to one chemist working with physicists to create new nanomaterials, it can take six months just to learn the physics of materials science well enough to help structure a simple experiment in materials creation.[4]

Finally, the most complex and sophisticated aspect of scientific capacity is the ability to retain knowledge for future access and use. Unless scientists can build on previous work, they are not likely to get very far in understanding the natural world. As Robert Hooke noted in his introduction to experimentation in 1666:

> There hath not been wanting in all ages and places great numbers of men whose genius and constitution hath inclined them to delight in the inquiry into the nature and causes of things, and from those inquirys to produce somewhat of use to themselves or mankind. But their Indeavours having been only single and scarce ever united, improved, or regulated by Art, have ended only in small inconsiderable product hardly worth naming.[5]

Without the ability to retain knowledge, whether within an institution, within publications or databases, or within individuals' understanding and practice, science cannot be useful or developed into usable bodies of knowledge.

Measuring Scientific Capacity: Constructing an Index

Scientific capacity is a complex and multifaceted concept that in some sense defies measurement. By collecting data on various inputs to scientific capacity and combining these data into an index, though, it is possible to get at least a rough sense of the degree to which a country is ready to enter or participate in the new invisible college.[6] The index presented here relies on eight key indicators: (1) per capita GDP, (2) the gross tertiary science enrollment ratio, (3) the number of scientists and engineers per million inhabitants, (4) the number of research institutions per million inhabitants, (5) research and development (R&D) spending as a share of GDP, (6) the number of patents per million inhabitants, (7) the number of articles published in international science and technology journals per million inhabitants, and (8) the comparative share of internationally coauthored papers held by each country in 2000. (See appendix A for details on the construction of the index.)

These indicators can be divided into three categories: "enabling factors" that help create an environment conducive to the absorption, retention, production, and diffusion of knowledge; "resources" that can be devoted directly to science and technology activities; and "embedded knowledge" of science and technology, including the extent to which researchers are connected to the global scientific community.

The enabling factors are represented by per capita GDP and the gross tertiary science enrollment ratio, or the number of students enrolled in higher education in the sciences as a share of the population that is old enough to have been out of high school for five years or fewer.[7] The first indicator indirectly measures the ease of working within a particular national system by serving as a proxy for information about roads, electricity, transportation, communications, and so on. The second measure shows the extent to which the nation supports, values, and provides an educated workforce. Even the largest investments in science are unlikely to pay dividends if a country's population remains largely untutored and unskilled. Trying to establish a center of, say, biomedical research in a country where illiteracy is rampant is like trying to plant an acorn in a desert. Projects are unlikely to take root, let alone thrive. To produce knowledge that delivers social benefits, research centers need more than just skilled technicians and support staff; they need to be connected to an economy with the ability to capture that knowledge and apply it to useful ends.

The resources available for science and technology are reflected in the number of scientists and engineers, the number of research institutions, and the amount of spending on R&D. These indicators measure the population's ability to engage in scientific problem solving, the extent to which scientists have access to research centers, and the financial resources the nation devotes to science as a whole.

The country's stock of embedded knowledge—which has been developed locally and could conceivably be tapped for future use—is represented by the number of patents and science and technology journal articles per million

Table 6-1. *A Ranking of Seventy-Six Countries, by Weighted Average*

Advanced countries		Developing countries		Developing countries (cont.)	
1	United States	28	Belarus	57	Malaysia
2	Canada	29	Portugal	58	Bolivia
3	Sweden	30	Slovakia	59	Tunisia
4	Finland	31	Hungary	60	Peru
5	Switzerland	32	Croatia	61	Bangladesh
6	Japan	33	Lithuania	62	Pakistan
7	Germany	34	Poland	63	Uganda
8	Israel	35	Bulgaria	64	Thailand
9	Australia	36	Cuba	65	Philippines
10	Denmark	37	Jordan	66	Egypt
11	United Kingdom	38	Argentina	67	Ecuador
12	Norway	39	Latvia		
13	France	40	Azerbaijan	*Lagging countries*	
14	Netherlands	41	Chile	68	Syria
15	Belgium	42	Macedonia	69	Senegal
16	Austria	43	Romania	70	Nicaragua
		44	South Africa	71	Indonesia
Proficient countries		45	Kazakhstan	72	Sri Lanka
17	Singapore	46	Moldova	73	Togo
18	Korea, Republic of	47	China	74	Central African
19	New Zealand	48	Kuwait		Republic
20	Ireland	49	Costa Rica	75	Nigeria
21	Russia	50	Brazil	76	Burkina Faso
22	Slovenia	51	Iran		
23	Italy	52	Turkey		
24	Spain	53	Mexico		
25	Estonia	54	Armenia		
26	Greece	55	India		
27	Czech Republic	56	Mauritius		

inhabitants. The country's comparative share of global papers measures its connectivity to the outside world, as well as the extent to which national researchers are working at a world-class level.

To construct an index, I standardize each of these indicators and combine them to produce a single measure of scientific capacity. I give resources, which represent direct measures of capacity, twice as much weight as enabling factors and embedded knowledge and connectivity. Table 6-1 presents the rank ordering of 76 countries that results from this process.

It is hardly surprising that the United States, Canada, Sweden, and other advanced industrialized countries, primarily in Europe, head the list. Indeed, if this were not the case, it might suggest that the index had been poorly constructed. But the table does include some surprises. For example, a few former Soviet bloc countries (Slovenia, Estonia, and the Czech Republic) rank among the scientifically proficient nations. Others, though (Hungary, Croatia, Lithuania, Poland, and Bulgaria), rank just above impoverished Cuba and Jordan. Brazil, Turkey, and Mexico all lag Azerbaijan. Breaking the index back into its major components can help explain these apparent anomalies; this question is discussed in appendix A.

It is important to emphasize that this index does not measure scientific capacity in absolute terms. It shows only how countries score on these indicators relative to each other. In addition, constraints on data availability limit the number of countries for which an overall measure can be calculated. Nonetheless, by adapting the approach underlying the index, policymakers can begin the task of inventorying the scientific capacity of their own countries.

Bringing Vietnam into the New Invisible College

Tran Ngoc Ca, deputy director of the Vietnamese National Institute for Science & Technology Policy and Strategy (NISTPASS), an independent division created within the country's Ministry of Science and Technology (MOST), began such an effort in 2000 when he undertook an inventory of Vietnam's S&T capacity. This was the starting point for creating a long-term plan to bring Vietnam into the community of top scientific nations. Officials at MOST had become convinced that the country needed to leverage science and technology more effectively to combat economic poverty and social deprivation. Accordingly, the small nation needed a bottom-up strategy to build scientific capacity for a knowledge-based economy. This was the challenge facing Ca.

BRAVE NEW WORLD

MOST authorities were beginning their quest to build scientific capacity in a world very different from the one described by Vannevar Bush in his post–World War II report, *Science, the Endless Frontier.*[8] During the era of scientific nationalism in the twentieth century, large governmental agencies were created in the developed world to manage science and technology and to foster innovation at the national level. The scientifically advanced countries instituted regulations, created standards, furnished funding, and built institutions to nurture and capture the benefits of science. But Vietnam could not hope to create such a system, given both its financial constraints and the vast increase in worldwide research. Officials at MOST recognized that Vietnam could not invest in all areas of science, nor could it build all the institutions needed to advance scientific capacity. To flourish, Vietnam's science strategy would have to create a system with an international focus. This would involve a big change for a nation that had endured decades of war followed by centralized planning and a period of relative isolation from the world.

Ca, with a doctorate in science and technology planning at the University of Edinburgh, did not need to be convinced of the wisdom of this approach. His studies, along with visits to Europe, North America, and Japan, had given him deep insight into the dynamics of the knowledge-based economy and the need to create a flexible, adaptive, and outward-looking system. "We have a debate in Vietnam," Ca explained. "Unless you choose the right investment it will be a waste of time and money. But what is the right investment? The answer goes back and forth between investing in infrastructure and investing in people. But I looked at it differently. To me, *isolation* is one of our biggest problems. We need to be more international."[9]

TAKING INVENTORY

Ca began his task by considering the enabling factors that support science and technology capacity. The initial outlook was not very bright. Vietnam's per capita GDP was very low, ranking 156th among all countries in 2002. Furthermore, the tertiary gross enrollment ratio across all fields was only 10 percent, according to the United Nations Educational, Scientific and Cultural Organization (UNESCO). But Ca was encouraged by the fact that Vietnam has been increasing its investment in education.[10]

When Ca turned his attention to the resources that could be devoted to science and technology, it became clear that although Vietnam had not invested much in science capacity, the country was not starting from zero. In the early 2000s Vietnam was home to more technically trained personnel than one would expect for a country of its size and level of economic development. The country's number of scientists and engineers per million inhabitants was close to the international median. Similarly, the number of research institutions in Vietnam, relative to its population, was about average for scientifically developing nations. But compared to its peers, Vietnam lagged in the percentage of its GDP that was devoted to R&D—the most commonly used measure of direct investment in science and technology—and a tiny portion of its national budget was committed to R&D. (Most funding came from donors.) The country had made some recent progress in this area, but pressing domestic demands made it difficult to increase investment in R&D.

Nonetheless, when Ca examined Vietnam's stock of embedded knowledge, he found that the country's scientific achievements exceeded expectations aligned with the size of its economy. In 2002, researchers based in Vietnam published close to 400 papers in internationally recognized science and technology journals. The country's performance was particularly strong in mathematics, which is often called "the language of science" and is closely related to the field of computer science. In addition, the country had strengths in materials science, including the development of nanostructures and polymers.

Ca also realized that less formally codified or even uncodified forms of embedded knowledge might play an equally important role in determining the country's science strategy. As he explained, "The real capacity is in people, and sometimes we call it by another name, but it is invested in people in firms, and universities. It is *learning*, and you can see various ways of learning: learning by doing, learning by formal training and education, learning through interaction."[11] Such learning and knowledge might give Vietnam an advantage in developing new knowledge and attracting the attention of researchers working abroad.

For example, Ca knew that shrimp farmers on the coast had a great deal of experience with managing aquaculture problems. The same story could be told of fruit farmers in the Mekong Delta. But because these groups were not always tied to research institutions, their unique knowledge of local conditions was unlikely to feed back into the research process. Therefore, policies had to be designed to capture this embedded knowledge, make it accessible to researchers, feed the results back to those working in the field, and obtain their feedback once again. In part this meant investing in new software, databases, and laboratory space, as well as in training for top students abroad. In addition, Vietnamese policymakers promoted links to international agricultural research centers, such as those sponsored by the Consultative Group on International Agricultural Research (an intergovernmental group hosted by the World Bank). These links granted access to important information about technical standards that was needed to advance local research. Policymakers also drew on the opportunities for local cooperation provided by the Asia Pacific Economic Cooperation (APEC) forum. For example, Vietnam and Japan are now cooperating on efforts to test the quality of local products.

This last example highlights the importance of connectivity in Vietnam's science policy. The question of how to link the country to a larger world community presented perhaps the greatest challenge. "The biggest problem we face comes in communicating with the outside world," Ca said. "We have good scientists working here—and they work hard—but they are often unknown to the larger world because they work and publish mostly in Vietnamese. We need to help our scientists become more confident and more connected internationally."

This problem might not have been immediately apparent from a quick look at measures of the country's connectedness to global science networks. On measures such as the comparative share of global papers originating from each country, Vietnam outperformed other countries. It had a very high rate of international collaboration, with more than 70 percent of all publications coauthored with scientists from other countries. But this is true for most small countries because they cannot offer the full range of resources that local scientists need. These facts raise two questions: Does collaboration show connectivity or dependence? Do the resources being shared at the global level help to build capacity at home? Connections to the international scientific community are sometimes formed at the expense of local connections, particularly when other parts of the scientific infrastructure do not exist or are weak. Scientific strengths also need to be tied down to local applications if they are to be sustainable over time.

To encourage both more interaction and more productive interaction, officials at MOST decided to let scientists take the lead in identifying opportunities for international collaboration. The ministry established an Internet chat room to involve scientists in virtual discussions on convexity and monotonicity (two important characteristics of some mathematic functions). Anyone who was interested could join this open forum. Senior scientists and students worked together and linked with scientists in other countries, many of whom had roots in Vietnam. "This way, we can organize subgroups that intensively discuss a topic," Ca explained. "When they get to the point where they want to meet face to face, they can apply for funding and show a history of interaction." This strategy soon bore fruit: based on connections established over the Internet, Vietnam hosted a world conference on convexity and monotonicity in 2002.[12]

SETTING PRIORITIES

These positive moves to identify and then build scientific capacity were only one part of Vietnam's strategy. MOST administrators also sought to identify priority areas for investment. The ministry could have approached this task in a variety of ways. Some nations, like the United States, set priorities using a political process that focuses on public missions of defense, energy, space, and health as a guide to investment. The U.S. process is heavily influenced by the legislature.

In Japan, priorities for science are determined through a consensual survey process conducted by the ministry tasked with supporting science and technology. That particular process collects feedback from hundreds of researchers and research managers on which sciences or technologies could help meet critical social needs.[13] Once collated, these views become a guide for budgeting for public science; the results heavily favor sciences that support industry.

In Europe, groups of experts and citizens meet to discuss social needs, scientific and technical opportunities, and available resources in a process known as "Foresight." The results of these deliberations inform the budget decisionmaking processes at both the national and the continental level. Although the influence of Foresight panels varies across countries, they are a common feature of European scientific priority setting. They also often have a social agenda.

Instead of adopting any of these methods, MOST officials opted to focus on identifying Vietnam's existing strengths, linking them to local needs and capabilities, and then building from there. The ministry realized that it would have to be selective in this process and make what were—in the Vietnamese context—sizable bets. According to Ca, the thinking that guided their efforts was "Don't spread resources too thin—be willing to take a risk on a big investment in a field where your country has both capabilities and needs." These investments should be made in a way that enables local institutions to learn from the process of meeting a locally relevant challenge. They should also involve both local investment capacity and connections to capabilities offered by regional or global partners.

MOST representatives identified Vietnam's general strengths—a culture that values knowledge and education, a large and low-cost labor pool, and goodwill from around the world, not least from the many expatriate Vietnamese who had trained as scientists and researchers in other countries. In addition, the ministry found that Vietnamese scientists appeared to be particularly successful in certain fields. In particular, articles by Vietnamese researchers on infectious diseases were more widely cited than articles in other fields. In addition, Vietnam offers many opportunities for local and international researchers to study diseases such as leprosy, avian influenza, severe acute respiratory syndrome (SARS), and pediatric dengue hemorrhagic fever. Among other issues, the emergence of drug-resistant strains of bacteria or viruses offers cause for concern and research. As a result, the ministry has chosen biotechnology as one area in which to focus Vietnam's scientific efforts in the twenty-first century.

Building Scientific Infrastructure

As the Vietnamese example shows, a basic inventory of the principal functions (summarized in Table 6-2) that contribute to scientific capacity can be a powerful starting point for formulating national science strategy. But if such a strategy is to succeed, policymakers must look beyond just funding science. A focus must include the functions and services that support scientific research and technological development. They must broaden their focus to encompass the infrastructure that supports scientific capacity. Without this basic scaffolding, attempts to build science capacity by focusing on research and development are doomed to crumble—knowledge-creating activities cannot be sustained without the services and functions of related science and technology services such as metrology or extension.

The encouraging news, however, is that even though science and technology infrastructure has traditionally been built at the national level, this is no longer necessary. Each country does not need to furnish by itself every function of this support system. Some components, such as standards setting and knowledge scanning, can be bought or borrowed from other places. These services can possibly be provided at the regional or international level by nongovernmental or intergovernmental organizations. An understanding of those basic components is the first step toward creating the infrastructure that will support a science and technology system.

Elements of a Science, Technology, and Engineering Infrastructure

Traditionally, this type of infrastructure has been defined as the physical plant and key services that contribute to the maintenance of a science and technology system. The primary components of a scientific infrastructure are (1) laboratories and equipment; (2) standards, testing, and metrology services, including regulatory and compliance services; (3) extension, technology transfer, and information collection services; and (4) intellectual property protection. Similar support functions can be found in all scientifically advanced countries, although from country to country they may be available from a dif-

Table 6-2. *The Functions of Scientific Capacity*

Infrastructure component or function	Subcomponent	Metric range in countries with advanced scientific and technical capabilities
Scientific and technical laboratories		Two to nine institutions per 100,000 inhabitants
	Laboratory equipment	R&D funds require an additional 20 percent spent on equipment
	Laboratory space	Between 250 and 1,000 square feet is allocated per research staff member
	Public spending on research	GERD spending of ~$60 million per 100,000 inhabitants[a]
	Government share of academic research	Government funds 60 to 70 percent of university-based research
	Government share of business research	Government funds more than 6 percent of business-based research
	Industrial contribution to academic research	Industry funds approximately 6 percent of academic research
Standards, testing, and metrology services		Governments spend $150,000 per 100,000 inhabitants on manufacturing extension services
Extension services, technology transfer, and information collection		Governments contribute close to $200,000 per 100,000 inhabitants on manufacturing extension services

Metrics for the following infrastructural components were not estimable with available data.

Intellectual property protection	Government provides legal framework; public sector grants and litigates intellectual property rights; patent offices often self-funded	Companies in many countries tend to patent in the United States, Europe, Japan, or all three to gain broad market protection
Vocational education and training	Government (national, regional, local), private sector offer training opportunities; spending varies considerably and is difficult to estimate	Governments are generally highly committed to vocational and technical training
Regulatory and compliance services	Government (national, regional, local) creates regulations and offers compliance services; businesses also offer compliance services; spending difficult to estimate	Different countries have a vastly different mix of regulations and services; the responsible party (public or private) also differs considerably among countries

Source: Author's calculations based on data from the Organization for Economic Cooperation and Development (OECD) and other government data.

a. Gross Expenditure on Research and Development (GERD) in 2005 U.S. dollars. Definition from OECD.

ferent mix of public, private, and academic sector institutions. In North America, for example, the public sector supplies most research funds. In Japan, though, the private sector funds the majority of research. Similarly, the public and private sectors play different roles in formulating and enforcing standards in Europe and the United States. This diversity of approaches reflects broad variation in the initial conditions from which scientific institutions emerged over time. As national systems evolved, each took on a different set of characteristics—testimony, incidentally, to the fact that the components of a complex adaptive system can organize and combine in many different ways. Here I do not try to sort out the question of whether any one system has advantages over another. Instead, I focus on the activities and functions that all working systems share. I also seek to evoke a sense of the magnitude and organization of these functions within the scientifically advanced countries.

SCIENTIFIC AND TECHNICAL INSTITUTIONS

These institutions are the backbone of any scientific infrastructure. Many countries see investments in laboratories and equipment (usually located within a university or independent research center) as a critical component of their science strategy. In addition, because institutions are usually made of bricks and mortar, they are easier to identify than less tangible assets like the skills and expertise of individuals. Items that are more visible are also easier to count, which means that more data are available about scientific and technical institutions and equipment than about any other element of the infrastructure. For example, data from the World Bank show that scientifically advanced countries have, on average, three scientific or technical institutions per 100,000 inhabitants.[14] This statistic is derived from an average among sixteen scientifically advanced nations. Within the United States, these institutions contain a total of more than 100 million square feet in research space, according to data collected by the National Science Foundation (NSF).[15] On average, leading researchers (principal investigators on funded grants) are allocated more than 2,000 square feet in research space each, compared to an allocation of more than 200 square feet for each individual working under their supervision.[16]

The government funds a significant share of the cost for the physical plant and equipment at these institutions. In 2000, the U.S. government contributed close to 10 percent of the total construction costs of academic research laboratories, even while covering the capital costs of its own laboratories.[17] In addition, the government often underwrites specialized equipment that researchers need, funded on a case-by-case basis. The NSF estimates that the federal government underwrote close to 60 percent of total spending on academic research instrumentation in 2000.[18]

Overall, in 2000, the average scientifically advanced country devoted more than 2 percent of its GDP to R&D. These funds are typically committed by and spread across three sectors: government laboratories, industrial laboratories, and academic laboratories. According to a report from the U.S. National Science Board, public spending on scientific and technological infrastructure in the United States amounted to 20 percent of total R&D spending for 2002.[19] Similarly, the NSF devotes more than 20 percent of its own budget to infrastructure. As a rule of thumb, then, it is possible to say that 20 cents of every research dollar goes to fund equipment and physical plants.[20] This figure can serve as a rough guide for policymakers seeking to determine how much should be allocated to science and technology funding.

STANDARDS, TESTING, AND METROLOGY

These functions are extremely important to advanced industrialized economies.[21] The scope and extent of these activities is vast, and the benefits that accrue to economies are wide-ranging. By using standards, testing, and metrology services, often supplied by a disinterested third party, firms can assure clients that products meet standards for quality, interoperability, or functionality. In science, for example, chemicals and microscopes must meet exacting standards to function according to the needs of scientific research.

Responsibility for setting and enforcing scientific standards varies across countries and fields. Some governments become directly involved in setting standards for both the public and private sectors; some stay out of setting standards in the private sector, getting involved only in determining their own specifications. In many cases, governments accept private standards into the public regime. In some cases, standards are set de facto in the marketplace or practice. In other cases, they are set de jure by a decisionmaking group. Such standards-setting bodies, which are often international in character, typically include individuals who represent both public and private bodies. For example, the Quantum Physics Division of the U.S. government's National Institute for Standards and Technology (NIST, an agency of the U.S. Department of Commerce) works with a committee of advisers on fundamental, highly accurate measurements and theoretical analyses, using quantum physics, quantum optics, chemical physics, gravitational physics, and geophysical measurements.[22] This includes, for example, developing the laser as a precise measurement tool. The technical standards developed are shared with any research body—foreign or domestic—interested in using these tools.

In all scientifically advanced countries, at least one institution is given responsibility for legal metrology, which is the regulation of weights and measures. Typically this is a national institute of measurement. In addition, accreditation institutions play an important role in these economies. These bodies, such as the Underwriters' Laboratories, offer third-party quality assurance guarantees for a fee. The services offered include calibration, testing, certification, inspection, and verification. These groups are often privately established and operated, although a number of international bodies exist to help ensure the consistency of accreditation procedures. They are available to any user, regardless of national origin.[23]

Extension, Technology Transfer, and Information Collection Services

Many countries sponsor extension and technology transfer services to aid research, development, testing, and evaluation. On average, scientifically advanced countries invest close to US$2 per capita on extension services.[24] These activities can take the form of "science shops" like those funded by the government of the Netherlands, which transfer knowledge from universities to industry. The government of Japan maintains a network of Kohsetsushi engineering centers throughout the country to aid industry with science applications, technology, and engineering adaptation.[25] Many countries support science and technology parks—offering low-cost land, loans for building construction, and tax breaks to companies that establish growth-based businesses within these centers. Other services offered include incubator centers that support small technology-based start-ups in Russia.[26]

Governments also collect and make available technical information about the capabilities of foreign research centers.[27] This type of intelligence gathering can be very valuable to users. Such services range widely in size and in the depth of information they offer. The services of the Japan Science and Technology Agency are perhaps the most highly developed. This agency collects, analyzes, and disseminates technical information from around the world for use in Japan. In a number of cases, the extension services work with the standards-setting or assurance bodies to ensure that local industries know and can certify their products to market standards.

Intellectual Property Protection

The question of who can own and control the results of scientific research is highly controversial and has given rise to numerous legal, administrative, and trade disputes.[28] Although the idea of knowledge as property is not new—it emerged around the same time as modern science—the scale and changing nature and structure of research have intensified the debate about the legal codification of this idea. This is particularly true with regard to biological and biotechnology products. For example, some foreign companies have patented and profitably exploited indigenous natural products, such as India's tea tree oil. Not surprisingly, the governments affected by such claims argue that such patents are inappropriate.[29] Similarly, the granting of patents for chimera and cloned biological materials has generated significant controversy and international acrimony.

Patents emerged originally from the idea that a time-limited monopoly would stimulate the creation, disclosure, and development of inventive or creative works. The importance of protecting the rights of inventors was written into the U.S. Constitution in 1789. Today, most scientifically advanced countries have an office of intellectual property protection that registers and enforces patents, copyrights, and trademarks. These services primarily benefit scientists in industry (and of course, their employers), who are more likely to patent than to publish their work. In the United States, however, patenting has also become increasingly important in the academic sector since the 1980 passage of the Bayh-Dole Act, which made it easier for universities to retain the rights to intellectual property resulting from government-sponsored research. Many universities have opened offices to identify and register intellectual property. As a result, more academic research now has commercial applications. But some critics believe that the price has been a reduction in the free flow of scientific information.[30] They argue that research that once might have been published is now often withheld pending the filing of an invention disclosure with the university and a patent application with the United States Patent and Trademark Office. In addition, much of that research might once have been freely available to interested parties, but it is now accessible only to those who can afford to pay for a license.

My previous discussion of the workings of the new invisible college suggests that any restrictions on the circulation of new ideas and data can reduce its effectiveness and productivity. Emergent knowledge creation depends on open networks of exchange. For this reason, scientific research should not be burdened by constraints, either political or financial. With that said, most countries do need an office that oversees the patent registration process for science- and technology-based industries, even if only to ensure access to materials and processes that might otherwise be held in monopoly by foreign investors. The broader goal of these offices, though, should be to work toward a more open intellectual property regime that will help bridge, not widen, the gap between the rich and the poor.

Beyond National Systems

As globalization deepens and expands its reach, it increasingly makes sense for nations to share the management of essential services like intellectual property protection and that these services be provided by coalitions and have the parameters negotiated at the international level. Such a system would do more to promote intellectual and economic exchange than the current structure of overlapping and competing national regimes. Countries should do more than just collaborate to fund and conduct scientific research; they should also cooperate in building the infrastructure that supports such activity. In cases such as metrology, in which efficiency is served by adherence to a single standard, governments may choose to join forces to offer services at the regional level. In other cases, such as information collection, it might make sense for a single entity to handle coordination at the global level, as the Global Biodiversity Information Facility (GBIF) does. Although having a blueprint for this division of labor is appealing, it will take experimentation and a trial-and-error process of matching capacities with local needs to find out what works best.

To summarize, the elements of a scientific support system can be combined and recombined in different ways. Any of the services shown in table 6-2 can be furnished by any of a range of institutions—public or private, international or domestic, commercial or not-for-profit—there is not one model that serves all nations. However, the functions and services are essential, and these parts of the system can be provided in many different ways. Many functions can be purchased, shared, or temporarily imported and do not need to be re-created locally. These questions can all be resolved separately from the question of what entities will fund these activities and who benefits—an issue addressed in the final chapter.

PART III

Tapping Networks to Extend the Benefits of Science and Technology

I would like to think that we are on our way to becoming an
embryonic central nervous system for the whole system.
I even like the notion that our cities, still primitive,
archaic, fragile structures, could turn into the
precursors of ganglia, to be ultimately linked
in a network around the planet.

Lewis Thomas,
"On the Uncertainty of Science,"
Harvard Magazine 83, no.1 (1980): 19–22.

CHAPTER SEVEN

GOVERNING THE NEW INVISIBLE COLLEGE

> Of all recent tendencies . . . the development of economic nationalism has been most dangerous to the application of science to human welfare.
>
> J. D. BERNAL, *The Social Function of Science*

S cientific nationalism defined the conduct and governance of science in the twentieth century, constraining the emergent organization that leads to the most efficient organization and creative outlets for scientific communications. The inefficiencies of cold-war science, in particular, hindered its application to human welfare as it became caught up in the power struggles of twentieth-century nations. This situation began to change only at the outset of the 1990s, as sweeping political change at the end of the cold war converged with the information revolution.

Indeed, in 1990, a new chapter in the history of the invisible college opened with the reintegration of scientists and engineers from the former Soviet Union into full communication with world science.[1] In Germany, Hans-Dieter Klenk, director of the University of Marburg Virology Institute, experienced this shift firsthand. At the 1990 International Congress of Virology in Berlin, Klenk met A. A. Chepurnov, a leading researcher at the State Research Center of Virology and Biotechnology VECTOR (the Vector Institute) in Russia, along with some of his colleagues. This was the first time the Russians had attended the conference. "We were very interested to hear what

J. D. Bernal, *The Social Function of Science* (London: G. Routledge & Sons, 1939), p.149.

they were working on," Klenk recalled. "When I met the Russian team from the Vector Institute and learned of their work, we realized that we could help each other."[2]

Chepurnov told Klenk that Russian researchers had initiated inquiries into hemorrhagic viruses in the 1980s, but that his laboratory had been forced to drop this work because it lacked the equipment needed to sequence the virus genome. When the two men met, Chepurnov realized that through collaboration with Klenk and others at Marburg, he had an opportunity to advance the promising work he had abandoned several years earlier. The Marburg Institute of Virology is world famous for having isolated and characterized the genetic structure of a hemorrhagic fever similar to the Ebola virus. Discovered in 1967, the fever was named Marburg virus after the German laboratory. Chepurnov was keen to work with Klenk and the Marburg team in hopes of furthering his research. Klenk, in turn, was interested in gaining access to Chepurnov's data on animal experimentation. Together, the two scientists began a research project to genetically characterize variants of the viruses that cause deadly hemorrhagic fever. In the process, they created a link in the new invisible college.

Within a year of completing the genetic research with the Russian team, Klenk and a group of his colleagues submitted an application to the World Intellectual Property Organization to patent a serum designed to treat the inflammation induced by hemorrhagic shock. By bringing basic research to bear on a specific health problem, this network of collaborators had developed a product that could help treat a strange, rare, and particularly virulent disease that wreaks devastation half a planet away in Gabon, Africa.

The experimentalists of the new invisible college—as exemplified by Klenk, Chepurnov, and the other scientists whose stories are told in this book—are the norm, not the exception, in science today. They self-organize into teams, share resources, and collaborate to solve scientific problems. Project teams dissolve when a collaborative project has achieved its goal. These teaming arrangements are not tied to place. When needed, research can be geographically distributed, flexible, and mobile. Nor are the teams limited by discipline or sectors: they tap into diverse fields of research as needed (in Klenk's case, crossing genomics, virology, epidemiology, and medicine), and work with private sector collaborators when the opportunities are attractive. Their work is nonlinear and complex, going from basic research to market applications and back again, all driven by a basic set of questions about the effects of and remediation for a virus on a human population.

This chapter summarizes the lessons for governance to be learned from the emergence of the invisible college as the dominant form of organization in science. Both the governance challenges and opportunities offered by the new invisible college differ significantly from the challenges and opportunities policymakers faced in the era of scientific nationalism. Because the emerging system of science is not national, policies based on national models will not yield the desired outcomes. Yet most policymakers still adhere to these national policies.

It is enticing to study the approaches to science that were successful in the recent past—such as the national innovation systems model that some economists have identified with the United States and Europe,[3] and the Asian Tigers model[4] that some have identified with South Korea—on the assumption that they are actionable models for developing nations today.[5] But scientific nationalism, and the related concept of a "national innovation system," while relevant in the twentieth century, are waning in relevance and will do little to help build scientific capacity in the developing world. Instead, contemporary policymakers must have a strategy for harnessing self-organizing scientific networks at the local, regional, and global levels outside of a "national systems" model. This is true even if the nation-state still articulates the policy for managing the knowledge system.

The networks that produce and disseminate knowledge operate both within and beyond the nation-state. In the early twenty-first century, the sheer *volume* of international networking in science is forcing a phase-shift from smaller, nationally based groupings of scientists to a single global network that is interconnected with every country of the world and includes many smaller clusters. This shift changes the locus of influence for directing science to the global level. As this book has shown, networks readily diffuse and extend knowledge. They are the backbone of twenty-first-century science. Attempting to control the flow of knowledge by limiting it to political borders constrains the very dynamics that make science so useful to humanity.

The networks that make up the new invisible college operate by clear, if not self-evident, rules. They grow from the bottom up rather than from the top down. Networks become complex as they grow and evolve. Their organization is driven by the forces and structures described in this book—preferential attachment and cumulative advantage, trust and social capital creation, and the incentive system that leads scientists to share data and exchange information. As a result, these networks cannot be managed; they can only be guided and influenced. They have more in common with the organic systems described in Sir John Evelyn's *Sylva* (see chapter 3) than with the clockwork of Newtonian calculus. To govern these systems, policymakers must understand their dynamics and then devise incentives that will lead individual scientists to make the decisions they want.

Accordingly, the key issues for science policy in the early twenty-first century are (1) how to create policies that take into account and encompass the many levels at which scientific networks operate, (2) how to align incentives to increase opportunities for local participation, and (3) how to accomplish these objectives in a way that democratizes decisionmaking about scientific investments and resource distribution. Two distinct governance implications emerge from this new perspective. First, lagging nations seeking to develop a science strategy have an unprecedented opportunity to take advantage of the emerging system. Second, existing national policies, particularly those in scientifically advanced countries, must be significantly reformed. A new governance model is needed, one that considers science as a global public good. This new model challenges policymakers to turn the axiom around and "think locally, act globally."

Science as a Global Public Good

Klenk and his team's hemorrhagic fever research project shows clearly how the new invisible college works. First, the mysteries of the natural phenomenon are fascinating and intriguing, attracting the attention of top scientists. Second, the research activities once clearly defined were distributed to those most able to conduct them effectively—regardless of where they were geographically located. Third, the integration of data and information occurred virtually as knowledge was produced by each part of the team. Finally, the resulting knowledge emerged from the integration of complementary capabilities on the part of the collaborators. It was only through brain circulation—a face-to-face meeting—that the research was initiated. This collaboration was path-dependent, building on previous work at Marburg, and it was shaped by the "gravitational pull" of specialized equipment in both Russia and Germany. The resulting knowledge was distributed among the project members, as well as to collaborators in Gabon, Africa. In Africa, it could be used to solve or anticipate real problems.

The Marburg case illustrates emergent collaboration and demonstrates how open access to equipment, data, and people enables knowledge to be created. These two principles are the natural outgrowth of the networked organization of scientific knowledge creation and the conditions under which it prospers. They embody the social norms that operate within science. The challenge for policymakers is to bring governance into accord with these principles and norms.

Like many other kinds of science, the virology research conducted by Klenk and his team was funded from government coffers with public funds. During the twentieth century, this became the prevailing form of funding for basic science.[6] The argument underlying public funding for science has been that scientific knowledge is a public good.[7] In economists' terms, knowledge is "nonrival" and "nonexcludable." In laypersons' terms, one person's consumption of the good does not diminish the quantity available to others, and once the good is available to one individual, it is essentially available to all. Others in the same group cannot be excluded from consuming the good or sharing in its benefits. For this reason, private agents are likely to underinvest in public goods, giving rise to the argument that the government should supply them. This is the case not only for basic scientific research, but also for such goods as education, law enforcement, and a clean environment.[8]

Unlike other public goods, though, such as law enforcement or a transportation infrastructure—which have a strong local use component—science does not necessarily benefit the place where it is produced, nor does it necessarily aid the people who pay for it. Scientific knowledge can be created in one place and its payoffs can be delivered to people in another place or in the future.[9] Outbreaks of Ebola and Marburg viruses, for example, have been reported in Africa in Sudan, Zaire, Gabon, and the two Congos: Republic of the Congo and Democratic Republic of the Congo. Research on such viruses conducted in Marburg, Germany, would most likely be applied to and aid people thousands of miles away. The results of this research would be highly unlikely to benefit European taxpayers directly. Yet Klenk and Chepurnov pursued their research because it presented an interesting scientific problem.

Scientific research can be conducted at the local, regional, or global level, depending on the scale and scope associated with the research. Its benefits can reach many people beyond those in a single political system (which is also why science and technology are compelling investments for philanthropists). It can be captured for private benefit; for example, when a researcher discovers and a company markets a new drug. The connection between supporting research and reaping its benefits, then, can be quite tenuous. Nonetheless, democratic governments continue to fund science because the social rate of return—the benefits to society that outweigh the cost—appears to be considerable.[10]

Creating and Absorbing Knowledge

If science is a global public good, if scientists themselves organize the most effective scientific networks, and if knowledge is diffused through networks, it follows that science policy should seek to support and encourage networks. Moving along this logical path, it becomes apparent that no nation can have a fully contained science system because all parts of science interact with and support each other. To create knowledge, scientists must find ways to identify and connect to each other. As a result, the goal of policy should be to create the most open and fluid system possible.

Moreover, if information can be created at various places and integrated into knowledge anywhere that expertise exists, then learning the process of integrating and absorbing knowledge at the local level emerges as a critical part of the process through which science can be used to solve problems. No single place will have all the institutions, services, and capacity needed to manage this process on its own. Resource sharing is critical to a successful science system. This conclusion also argues for an open system that enables the most efficient sharing of resources.

Given all these elements of interconnection, policy in the twenty-first century should be driven by a vision of an open, interdependent, evolutionary knowledge system. Two key principles—open funding and open access to scientific resources and results—have the greatest chance of creating such a knowledge system. These principles can be achieved by crafting incentives that encourage and make use of preferential attachment to produce the most efficient organization of researchers; by focusing on bringing knowledge from anywhere in the world to bear on local problems; and by sustaining excellent science by applying it to the most pressing local problems, thereby ensuring local learning, feedback, and reward.

If the best available knowledge is needed to solve a problem or help a government meet a mission, it makes sense to fund the research of the team that conducts the highest quality science and has the greatest chance of applying it locally, without regard to the national affiliations of the team members. Any group that offers funding for scientific research should extend that offer to the most worthy team, and incentives can be proffered to ensure that the research addresses local or national problems. This ensures that scientific funds are being spent efficiently, rather than being channeled into less efficient spending for political purposes.

As part of an open system, governments could fund science through nonpolitical citizen councils, which would coordinate with existing scientific academies and agencies to design visions for the role of science at the local, regional, and global levels. All funding would be open, in the sense that any group or organization could compete for funds. A database of inquiries would be made available to help researchers find collaborators.

Greater citizen involvement could also increase the relevance of science to public needs by targeting research funds more effectively on the issues, problems, and opportunities of greatest local concern. Models of citizen involvement in scientific decisionmaking, including Foresight and Futura in Europe, could be applied in other places. In addition, a number of developing countries have created their own processes for gathering local input. Uganda is a leading example. As Judi Wakhungu explains, officials working under the aegis of the Uganda National Council for Science and Technology (UNCST) have involved community development officers, as well as the representatives of national research institutions, in the science policy process. The goal is to ensure that the interests of all stakeholders are addressed:

> Not only have they done that in terms of biotechnology, they have also done that in various other sectors, for example water, fisheries, wildlife conservation, and also the drafting of the fisheries. Uganda has even gone further to make sure that, in terms of handling and understanding and getting the message across in a balanced way about biotechnology for the citizens to be able to participate effectively [in] many of these, a lot of the information has actually been translated into the local languages, and so far I think Uganda is the only African country that has gone that far.[11]

The open model fits with the natural structure of science, and it has the greatest promise to diffuse the benefits of science broadly and equitably. This model, however, is not in use now, and it would be foolish to say that the transition from the current national system to an open non-national system will be easy. Tension will always exist between the needs of the political process and the growth of the knowledge network. As long as the public treasury is used to fund science, public needs will influence, if not define, the scientific agenda, appropriately so in many cases. The challenge for policymakers is to strike the balance between allowing emergent properties of science to flourish and finding ways to use the system to meet national goals.

New Approaches to Governance

Using the principles of open funding and open access as guides, we can imagine a new framework for governing science in the twenty-first century. The new approach uncouples science from national prestige and ties it more firmly to collaboration, merit, and openness; it makes research dependent on the needs of science instead of the interests of funders. Where the old system was nationally based, the emerging system looks from the local to the global level. Where the old system focused on institutional affiliation and structure, the emerging system focuses on the functions needed to facilitate knowledge creation and absorption. Where the old system was centered on strategic investments for competitive advantage, the emerging system focuses on cooperation for knowledge integration. Where the old system was focused on producing knowledge, the emerging system concentrates on absorbing and using it. Where the measure of the old system was inputs, the measure of the emerging system is outputs, often expressed in terms of social welfare. Where the old system protected national interests, the emerging system is open and fluid among countries.

These shifts require policymakers to adopt a two-part strategy. They need both an investment or "sinking" strategy and a communications or "linking" strategy. The contours and scope of these strategies might differ significantly among countries, depending on their scientific capacity and infrastructure. Table 7-1 presents some of the possible elements of these strategies and the functions they would serve.

Regardless of the level of scientific capacity from which a region begins, a sinking strategy should be based on the scale (the initial cost of establishing local capacity) and the scope (the long-term investment needed for sustainability) required to allow knowledge to be developed and locally absorbed. Investment in R&D capabilities may or may not be necessary for local absorption of knowledge. In some cases, local capacity can be replaced by a linking strategy. Decisionmaking about sinking can also require links across sectors, such as connections among private and public research groups. These factors can be determined in a number of ways, including an assessment of the scale and scope of similar investments in other places. The investment strategy should be created without regard to geographic borders and should consider the possible investment that can be drawn from local, regional, and global links. A workable strategy should consider brick-and-mortar investments and educational needs as well as communications investments.

The linking strategy should be understood as building a network at whatever scale is required for scientific research and knowledge access and diffusion. Using a network model allows policymakers to ignore the political boundaries that have defined national systems. Relying on such a model also enables planners to construct knowledge-creating teams and to gain access to information that may not be locally available. Much of the knowledge

Table 7-1. *Policy Steps in Linking and Sinking Strategies*

Function	Linking strategy	Sinking strategy
Assess local, national, and regional capacities, opportunities, and problems on a spectrum beginning with acute/local problems and moving to chronic/global problems (see table 7-2).	For those fields of science where capacity is lagging, but where there is interest in monitoring or drawing on knowledge in the field, create a linking post in a university with the greatest potential to absorb the knowledge and diffuse it.	For those fields of science where local, national, or regional capacity is needed, assess the pattern of geographic investment in other places. Assess and price infrastructure needs over the long term.
Scan the existing global knowledge base for useful information and individuals, assess the context of the available knowledge, and detail the "unknowns" associated with knowledge as it relates to specific problems and opportunities.	Collect or create reports on knowledge centers and capacities around the world. Scan the field for opportunities to collaborate or cooperate with different centers. Scope out research directions; identify fruitful directions that can be part of an investment strategy.	For the unknowns related to specific problems or challenges, where local capacity is needed, develop a collaboration strategy to link to other centers conducting similar or complementary research.
Determine the scale and scope of investment required to make existing knowledge locally accessible and applicable in a sustainable way.	Draw lessons on scale and scope from interactions with other research centers. Develop ongoing communications or participation in a consortium to aid local capacity if necessary. Consider the option of long-distance training.	For those fields determined to be critical in addressing chronic local or regional challenges, develop an investment strategy that accounts for infrastructure, training, and salaries.
Determine the scale and scope of investment needed to close the gap between the unknowns and the local and regional needs.	In fields where a linking strategy alone will be used, use the Internet services to alert participants to new developments; attendance at symposia or conferences is needed.	For those fields where investments will be made, focus on leading-edge research that can take advantage of local or regional feedback and learning. Feed this back into the linking strategy.
Identify possible partnerships and collaborations to make the required investment or otherwise access useful knowledge.	For those fields where a linking strategy is to be implemented, develop a range of collaborations, from periodic information sharing to research projects to workshops or conferences.	When investments will be made, put specific plans in place to enable collaborative research that enhances efficiency and complementarity, shares resources, and trains students.
Develop a financial and capacity-building plan to sink investments locally and link to regional and global capabilities where they can help.	Secure on-the-ground capacities that can absorb knowledge and establish the basis for scaling up if necessary. Local and regional capacities may differ in size, but will still need trained people and broadband communications tools.	Sink local or regional investments on a scale that accommodates leading-edge research needs; maintain investment over time. Links with local business or other academic centers are often critical.

needed to create local or regional solutions may be available through communications technology links or through collaborations that do not require significant brick-and-mortar investments. Thinking of science or engineering as a network also has the advantage of allowing a broader discussion for an action agenda. For example, in some cases, the communications strategy might include "outsourcing" parts of scientific infrastructure functions (such as standardization) to an established provider.

Both sinking and linking strategies need to serve a number of critical functions that require periodic review and reassessment, as shown in table 7-1. The first step in determining investment and allocation of funds for science is to *assess capacities* locally and define opportunities and problems on a spectrum beginning with acute (requiring immediate action) local problems and then moving to acute global problems, chronic (requiring the building of long-term capacity) local problems, and chronic global problems. This step can begin with an inventory or assessment of capacities based on laboratory space, publications, trained people, and so on, as outlined in chapter 6. This should be joined to an inventory of the problems to which science or engineering can be applied. Obviously, not all problems or challenges are amenable to a technical solution. But in cases of environment, agriculture, health, and basic industries, scientific resources or engineering skills can be applied in many cases.

An inventory should assign challenges to a category based on whether they are acute or chronic and local or global (see table 7-2). For acute challenges and problems, solutions or remediation should fall to *any* scientific or engineering team available anywhere in the world that can help immediately. No one would expect a scientifically lagging country to handle an outbreak of hemorrhagic fever on its own. Acute problems already fall within the purview of nongovernmental organizations (NGOs) like the World Health Organization. Perhaps this should be formalized into a global science corps that can be deployed to aid poorer countries as needed. For many poor countries, the solution to problems should remain in country, if only while they build a more robust scientific or engineering capacity.

Chronic problems or challenges should be the subject of additional analysis and strategic planning. Local or regional challenges should be among the first targets for assessment, so questions like how to meet the UN Millennium Development Goals of providing clean water, improving maternal health, remediating local pollution, improving soil conditions, and increasing the productivity of aquaculture should be on the agenda.[12] To the extent that they are not being addressed by existing programs, they should be part of an

Table 7-2. *Typology of Challenges*

Scale of impact	Type of challenge	
	Acute	*Chronic*
Local	Single water-pollution event	Poor soil conditions
Global	Viral epidemic	Energy availability

initial plan. For each of these areas, key sciences or technologies can be identified, scaled, and mapped at various levels.

The second step in a science planning process should be to *scan the existing global knowledge base* for useful information, centers, and people. This information should be used to create a virtual map of science for important fields. Policymakers should assess the context of the available knowledge to see what can be borrowed and what must be locally built. In other words, they should identify the extent to which research can tap available data (such as large-scale, online databases) or uses specialized equipment that can be accessed virtually. The location of this equipment is clearly important for a visual mapping of the field, which should be an integral part of the process. A list of key researchers and centers of excellence should be compiled to identify highly attractive people, places, and equipment to link to within each field. The goals are to leverage the equipment or capacities that exist elsewhere and to find resources that local scientists can access when they need them. Completing a detailed list of the key unknowns relating to specific problems and opportunities—or the most important questions researchers should address—should also be part of this process.

Science policymakers should then *determine the scale and scope of investment* required to make existing knowledge locally accessible and applicable in a sustainable way. A targeted strategy should be developed for challenges that meet four criteria: (1) they are amenable to scientific research, (2) they are chronic or recurring, (3) they are critical to problems, issues, or challenges determined to have local or regional applicability, and (4) they can support scalable research at the local level. This strategy will include some on-the-ground infrastructure and institutional investment as well as a plan to link into existing research in collaborative projects. In many cases, the scale and scope of the research will not be suited to investment by one nation. A regional or international coalition might be necessary to achieve adequate scale.

As part of this targeted strategy, it is important to determine the scale and scope of the investment needed to close the gap between the unknowns and local and regional needs or, in other words, to conduct leading-edge, exploratory research that is locally relevant. In many cases, exploratory research that is applicable and relevant to local problems and challenges will not be available from researchers elsewhere in the world. The answers to aquaculture diseases or regional plant blights, for example, may be highly localized. To the extent that a priority research plan can make these problems scientifically attractive to outside researchers, it may be possible to attract global collaborators and funding to local problems, increasing the productivity of research or bringing additional networked resources to bear.

These assessments and characterizations of the existing capacity, challenges, and global resources should be wrapped into a long-term strategy that *creates a financial and capacity-building plan* to both sink investments locally and link to regional and global capabilities where they can help. These plans can include long-term strategies for building capacity by training scientists, making capital investments, linking with donors and others who are funding related challenges, and establishing a targeted collaboration strategy.

Policy for Scientifically Developing Countries

This vision of science beyond the nation-state is one where local, national, and regional authorities have the capacity to do the science needed to solve real problems. But that capacity does not need to be available locally to be effective. In the future, no country will be able to make the kind of full-scale investment in science that the United States, western European countries, and the former Soviet Union made during the era of scientific nationalism. As the frontiers of science continue to expand, *all* science will move toward a process of sharing resources and results. As poorer countries solve problems, they will build capacities that will in turn be available to the scientifically advanced countries.

Fortunately for governments of developing countries, there is no longer an argument in the donor and development communities about whether investment in science and technology is worthwhile—this has become an accepted fact. Recent reports by donor agencies, advanced country governments, and nongovernmental associations have highlighted the importance of science and technology for development. The World Bank established an office of science and technology for development in the early 2000s, and the UN has made science a priority for development.

Even though participation in the global economy depends to some extent on national investment in science, the structure of this investment should reflect the structure of science. The capacity and infrastructure of scientifically developing countries do not need to—nor should they—mimic those of the scientifically advanced countries. Capacity and infrastructure can be created in collaborative multinational teams. Infrastructure can be accessed virtually through links to larger laboratories in other countries. Standards processes can be outsourced to existing institutions; for example, perhaps a large international scientific institution could become a broker of standards services. Expatriate researchers who work abroad can be tapped to help their native countries.

Indeed, developing countries have an advantage over developed countries—they have not built a twentieth-century national science system. This may seem counterintuitive; after all, most developing countries want to have highly developed scientific capabilities. But because these countries do not have the embedded twentieth-century bureaucracies and institutions that were the hallmarks of the era of scientific nationalism, they have greater flexibility to pursue new developments in science. The absence of nationally driven constraints tied to a huge investment can actually be an advantage that developing countries can exploit by building a more nimble networked system.

The issues facing scientifically developing countries are 1) at what level to invest and 2) how to select priorities for science and technology investment. The solutions will depend both on national needs and on the nature of individual fields. For those scientific research areas with more "stickiness," the strategy should include a global scan that identifies immovable resources to which local scientists can link. Within scientifically advanced countries, the incentives to open research facilities to others may be weak because of the vestiges of scientific nationalism. In this area, NGOs could help to negotiate increasing openness to researchers from scientifically lagging countries.

For the "lighter" fields, in which location is not as critical to research advances, the strategy should account for the possibility that local, regional, and global factors may influence investment decisions. In the field of mathematics, for example, there is some need for local capacity to absorb global knowledge and "tie it down," but the scale and scope of the investment in that capacity need not be large. Indeed, Vietnam built such a capacity in mathematics with very little up-front investment, in part by linking existing resources and people together.

Decisions about infrastructure investments can then be made either within national boundaries or at the local or regional levels. The scale and scope of the research determine the appropriate choice. Researchers in high-energy physics need only a few synchrotrons around the world. Agricultural researchers need locally available laboratories that can adapt to local conditions. These kinds of infrastructure choices will require that decisions be made on the basis of local needs and capabilities by field.

A global ministry of science is not the solution to these challenges. Even national ministries of science are suspect; broad institutions that seek to manage science funding may overly burden the network. Networks do not need a master plan. Indeed, such plans stifle them. To create robust networks and thereby distribute scientific capacities more broadly and fairly, incentives, resources, space for interaction, open funding, shared ideas, and feedback loops will lock in benefits. And these principles should underlie any investment strategy.

Nations still have sovereign interests, and accounting for the benefits of investment will remain an important feature of national policy. In some cases, nations will have historical precedents that will make crafting regional, geographically proximate collaborations difficult. In other cases, poor political governance or corruption will get in the way of fashioning a science policy for the twenty-first century. In still other cases, the large size of a national system may greatly complicate efforts to create a flexible and open global policy; ironically, this may be most true of the United States. Enlightened organizations such as philanthropic foundations may need to take the lead in promoting greater openness in science and technology and in working with lagging countries to make progress in applying knowledge at the local level.

Policy for Scientifically Advanced Nations

Scientifically advanced nations face three challenges in transitioning to the world of the new invisible college. The first and most demanding is moving from a perspective that views international collaboration as "external affairs" or "international relations" toward one that views the global system of science as the norm. The second challenge is redefining the role of these nations so that they see themselves not as donors of knowledge, but as participants in a complex system for creating and sharing knowledge in which they both absorb and contribute resources. The third is working with many different groups to develop the concepts and instruments needed to overcome the political obstacles that hinder collaboration and self-organization in science.

Given their embedded investment in scientific nationalism, advanced countries can be expected to be less nimble than scientifically developing regions in negotiating this transition. Changing the mix of investments for these countries is difficult—and has been likened to turning an aircraft carrier—because of inertia embedded in existing institutions and ministries. Yet advanced countries do have an important interest in cooperation and collaboration. Not only does cooperation open access to unique resources, such as the soil in Brazil for Wolfgang Wilcke, but it can also bring about new ways of thinking about a problem, thereby enhancing creativity.

Scientifically advanced countries also need linking and sinking strategies because no nation, however wealthy, will be able to invest in all frontier areas of science and technology. All nations will have increasing incentives to cooperate, even if only to reduce costs. To the extent that scientific infrastructure can be shared and not built redundantly in country after country (as was the case during the era of scientific nationalism), all science will become more efficient. Clearly, in some cases, local access to critical institutions and facilities is needed to maintain scientific capacity. In other cases, linking to the global network may reveal the necessary resources. These decisions will need to be made for each field and region, depending on the resources, challenges, and needs for feedback associated with particular learning processes.

Some investments already fall along the lines recommended here, such as those related to megascience projects. These cooperative activities are funded by multiple countries that band together because the costs are too great for any one country to bear or because the hoped-for benefits lie far in the future. Space research and fusion research have given rise to such collaborations. In addition, physicists working in these and related areas have been in the forefront of efforts to share research results with the widest possible group. As a regular practice, physicists put data on the World Wide Web for anyone to use. Many also publish their work on an open online forum.[13]

In contrast, most scientifically advanced countries lag when it comes to developing a strategy for distributed, bottom-up science. As discussed in earlier chapters, geographically distributed collaboration is growing faster than equipment- and resource-based collaboration, in large part because the Internet has reduced the transaction costs associated with distributed work. The Human Genome Project is one high-profile example of task-sharing across geographic space, with six countries sharing tasks and data in a distributed format.[14] As distributed collaboration increases, the question becomes one of how to access knowledge created in laboratories in faraway places. Accepting that geographic proximity matters in some fields of science or at some point in the research process will lead policymakers and researchers to work together, using virtual communication to reduce geographic distance. If critical knowledge is being created in a distant place, gaining the experience needed to absorb knowledge locally becomes the overriding challenge.

Essential first steps include identifying and mapping the locations all over the world where good research is taking place and helping scientists access this information. The Japan Information Center for Science and Technology, which is funded by the country's government, has provided such information for years to both the public and the scientific community. Other governments or NGOs should consider taking on a similar role in disseminating information about the scientific landscape.

In addition, to plan effectively for the future, policymakers must promote public understanding of science and technology and participation in the science policy process. To the extent that the world economies continue to grow toward knowledge-based societies, and to the extent that knowledge-based societies rely on science and technology, it is essential that the public gain understanding of science and technology. Further, public input to decisionmaking—beyond simply making market-based choices for technological products—is critical. Social tensions around the pace of change can be highly disruptive. Public participation in decisionmaking about scientific investments could potentially ameliorate some of these tensions.

Ultimately, what is needed is a policy that focuses not on scientific competition with other nations but on scientific collaboration and cooperation. Allowing political motives to determine the subject or object of cooperation is a temptation that governments should avoid. Commitments to set aside funds to cooperate with a single nation (whether the science is useful or not) may be politically expedient but are scientifically inefficient. Binational collaboration treaties should be avoided. Instead, science should be funded to meet scientific and social goals, not political ones.

Guiding Networks

If they are to participate effectively in the new invisible college, policymakers in both developing and developed countries must learn how to manage and govern emergent networks. These networks cannot be controlled; they can only be guided. Recall that networks evolve continuously based on the needs of network members and the incentives offered and available to network members to join and remain within the group. In the case of science, these needs and incentives most often revolve around the desire for recognition in its broadest sense. In consequence, the process of preferential attachment—who is most likely to help solve a specific problem or reach a particular goal—shapes the growth of the new invisible college. As we have seen, the most highly connected people are visible and "rich" with resources and therefore increase their connectivity faster than their less connected peers. This allows the influential hubs to consolidate their position as attractive points or centers of traffic and exchange within the network. Their position gives them influence over future connections and exchanges, making them disproportionately powerful in shaping research processes and outcomes.

The following guidelines derive from the network theory offered in this book. To promote the productivity of global science and encourage researchers to self-organize around questions of local, national, or regional concern, policymakers should

—Invite "champions" or highly influential scientists to help organize or lead research. Champions often act as "gatekeepers" to scientific resources, and can include new researchers.

—Facilitate interaction among many actors, particularly through face-to-face meetings at symposia and conferences as well as through funding for international travel and short-term studies.

—Create incentives to organize around interesting research questions. Calibrate these incentives based on "market" feedback as researchers seek to conduct research.

—Establish goals for research that are in keeping with spillover potentials at the *local* level.

—Create conditions for sharing knowledge, ideas, data, and codified information. Options can include grid computing, opening Web portals, scanning for important information, and creating projects, such as the Human Frontiers Science Program, that are established and funded specifically for international collaboration.

—Enable teams to outline rules of interaction rather than allowing these rules of engagement to be predetermined by an agency or institution. Guidelines can be helpful, but each team needs to establish its own rules for collaboration, for managing intellectual property, and for publishing information.

—Provide information about the landscape of science at the local, national, regional, and global levels so that local scientists know what is happening at the many venues where research is taking place.

The Role of Organizations and Institutions in the Networked Century

Networks do not replace organizations and institutions, but they do change the way they operate. As others have noted, we have a relatively good understanding of how to create institutions that are rule-bound, accountable, and reasonably effective in the vertical silos we call disciplines, the established institutions and the geographically bound nation-states. But we do not have adequate institutions or the ability to command horizontal accountability across states or even across academic disciplines. The question before us is whether networks can replace some of the roles we would like horizontal institutions to play. This has been the experience, for example, with the World Bank's CGIAR (Consultative Group on International Agricultural Research), which operates as a network with considerable autonomy within the parent institution. Networks offer a kind of flexibility and adaptability that institutions cannot; in cases where a flexible structure and adaptability are necessary, a network structure may be preferable to creating a new institution.

Funding is the main issue dogging change, and it is an issue where a reasonable solution will take time. We cannot change overnight the political links between science and national public funding, and in any case, perhaps we do not want to change too quickly. A democratic populace is served by science, and thus, scientific practitioners still need to be accountable to the public. This is currently done through nation-states, and is only changing slowly to a broader constituency. The public needs to understand the benefits of treating science as a global resource that can be locally tapped, and as this awareness grows, governing structures will find new ways to shift accountability to local benefits. To do this, policymakers will need to find new ways to measure the returns to science based on its global reach. As this happens, the logic of open funding of science will become more acceptable and viewed as inevitable.

Other models for funding need to be considered and developed in the meantime. The open-source software movement may provide such a model. There is no central funding for open-source software; funding is provided by and to individual network participants. Questions of intellectual property ownership are negotiated within the network. Knowledge is shared with all, encouraging creativity and additional sharing in a virtuous cycle. As a result, the productivity of the open-source movement has been immense. Similarly, discussions about funding for the publication of refereed scientific literature have focused on shifting from having the reader pay for access to having the author pay for publication.

As such new models emerge, nation-to-nation treaties and high-level agreements about science should be allowed to fade away. Science should not be the object of political trading or favoritism. International competition for scientific resources can only diminish all of science. Instead, policymakers around the world should form coalitions charged with addressing leading-edge scientific and technical problems and jointly creating incentives to solve these problems. When scientists everywhere step up to these grand challenges, policymakers should step away and let the new invisible college do its work.

Measuring Science and Technology Capacity at the National Level

A country's ability to participate in the global knowledge economy and in collaborative research at the international level depends largely on its capabilities in science and technology (S&T).[1] In this appendix, I present the classification scheme we used to measure S&T capacity at the national level.

For the purpose of this exercise, S&T capacity is defined as the ability to absorb and retain specialized knowledge and to exploit it to conduct research, meet needs, and develop efficient products and processes. S&T *capacity* is distinct from S&T *outcomes*. The question of which countries produce more scientific output is interesting in its own right. It can be addressed by ranking countries based on indicators such as the number of scientific papers or patents published or assigned each year. But many countries would be absent from this ranking. And such a ranking would also offer little insight into the ability of developing countries to expand their S&T capabilities in the future, to join international collaborations, or to use existing resources to build additional capacity. Yet these are the countries that most want to join the international S&T community. Accordingly, an index focusing on capacity can yield insights that could not be gained by examining outputs alone.

This work was done in collaboration with Edwin Horlings, Rathenau Instituut, The Hague, Netherlands, and Arindam Dutta, RAND Corporation, Santa Monica, Calif.

The index that I describe here—called the Science & Technology Capacity Index-2002 (STCI-02)—compares S&T infrastructure and knowledge-absorptive capabilities across countries.[2] It does not measure the extent to which a country is advancing the frontiers of S&T knowledge or producing S&T products. A high measure of capacity does not mean that scientific activity is actually being conducted in a country, but only that the conditions for such activity are present. Consequently, a country with little scientific output, such as New Zealand, may appear relatively high on the index because it has significant S&T capacity, even though its economic system is not currently exploiting this potential. By focusing on the conditions underlying capacity, this index takes a complementary approach to measurement efforts that focus on technological output, such as the Technology Achievement Index and the ArCo Index.[3]

Equally important, the index measures only *comparative* international differences and cannot be used to track the evolution of one nation's capacity over time. Moreover, focusing on capacity at the national level can paint only a partial picture of how S&T potential is distributed around the world. We use "country" or "nation" as the defining unit of analysis largely because this is how data are collected and reported. The nation-state provides a system of reference.[4] In addition, research and development (R&D) funds are often allocated at the national level and distributed through nationally based institutions. But knowledge does not honor political borders. Even within specific countries, the "knowledge border" can involve subregions as well as regions that cross national borders.[5] Within North America, for example, some subnational innovation systems are quite strong. In the European Union (EU), the S&T system is quite well integrated at the supra-national level.[6] More broadly, as this book argues, the emerging global knowledge system is changing the way the national systems operate.

Choosing Indicators

In choosing the indicators for the STCI-02, we acknowledge that S&T capacity is a theoretical construct. Its magnitude cannot be determined directly, let alone precisely. Instead, we must rely on a series of proxy variables. The indicators chosen for the STCI-02, then, are based on an assessment of the factors that enable countries to absorb, retain, use, and create knowledge. As a result, some of the measures overlap in terms of the features being assessed.

We took most of our data from published sources, such as the UNDP's human development reports. Collecting the data directly would have allowed us to adjust the definition and quality of the data to the needs of our exercise. Given the expense and time that this endeavor would have required, however, we chose to rely on the statistical publications of major international organizations. These are trusted sources of information with a long record in the collection and harmonization of data.

With these conditions in mind, we selected eight quantitative indicators:

—Per capita GDP in purchasing power parity dollars[7]

—Gross tertiary science enrollment ratio[8]

—Number of scientists and engineers engaged in R&D per million inhabitants[9]

—Number of research institutions per million inhabitants[10]

—Funds spent on research and development by public and private sources as a percentage of GDP[11]

—Number of patents per million inhabitants[12]

—Number of S&T journal articles per million inhabitants[13]

—Each country's weighted share of all internationally coauthored papers.[14]

As noted in chapter 6, these indicators are divided into three categories. Per capita GDP and the gross tertiary science enrollment ratio are examples of *enabling factors;* the number of scientists and engineers, the number of research institutions, and the amount of spending on R&D represent *resources* available for S&T; and the number of patents and S&T journals, as well as the country's comparative share of internationally coauthored papers, measures *embedded knowledge.*

In selecting these indicators, we had to strike a balance between coverage (number of countries, regions, or other units included in the analysis) and comprehensiveness (number and variety of S&T dimensions). Obviously, because no index can cover the entire range of dimensions related to a subject area, it is advisable to sharply focus the index—that is, to define its purpose and to outline what it does and does not measure. Coverage is intimately related to comprehensiveness: the more detailed the data will have to be, the fewer countries, regions, or social groups can be included. This is especially true of many developing countries where fewer data are collected and statistical information is often less reliable.[15] But any wide international comparison across a range of variables inevitably leads to problems with data availability. Accordingly, we sought to construct the index so we could include as many countries as possible, but the final list does not cover the entire world.

Table A-1. *Summary of Data Coverage of the Countries Counted in the S&T Capacity Index*[a]

Item	Number of indicators covered by global data search							
	All eight indicators found	Seven	Six	Five	Four	Three	Two	One or no indicators available
Number of countries with this level of data	**76**	19	25	22	4	5	31	1
Average per capita GDP ($US2000)	**8,329**	5,094	1,763	3,006	13,236	1,748	3,386	n.a.
Share of world population reflected in data	**82.4**	3.3	8.7	3.7	0.5	0.5	0.9	0
Share of world GDP reflected in data	**92.7**	2.2	2.1	1.5	0.9	0.1	0.4	0.1

n.a. = Not available.

a. Only countries with all eight indicators are included in the index; these countries are represented in the data given in the column in bold type. Base year is 2000.

Finding the balance between coverage and comprehensiveness involves choosing among three options: (1) using fewer variables, (2) reducing the sample of countries, and (3) devising statistical methods to preserve both the entire sample of countries and the entire set of variables.

We chose the second option after determining that we would sacrifice more by eliminating variables than we would gain in breadth of coverage. This choice may introduce a bias in the STCI-02 toward the highly developed countries. On the other hand, as data become thinner, we assume that the chances of missing comparative S&T capacity also diminish. This is suggested in table A-1, where the data coverage is shown.

We started with a list of 215 countries and territories based on international statistical publications. We excluded 32 dependencies and small nations for which data were extremely scarce, such as Andorra and San Marino, along with Tuvalu, Tonga, and other Pacific island nations, leaving a list of 183 countries. We then excluded countries for which data on one or more of the eight indicators were not available, as shown in table A-1. The 76 remaining countries—for which we have enough data to calculate the index—represent more than 80 percent of the world's population.

We addressed the possibility that relying on this subset would bias the index by constructing Lorenz curves of the income distribution of the countries in our sample and of all 215 countries and dependencies in the world. This analysis showed that the distributions are almost identical, although our sample has a higher average per capita GDP. A slight bias does result, but it does not affect the validity or representativeness of the index.

Constructing the Index

After selecting the variables, we combined the indicators into a single index by converting them into a common format, checking their consistency and correlation, and choosing a theoretically appropriate weighting scheme. This section presents our data and analysis in a way that allows other researchers to replicate our calculations and to revise the index with new information, adding or omitting variables as desired.

Conversion into a Common Format

To be usable, indicators must be comparable. For example, in financial indices, the values expressed are in the same currency (such as dollars). Data expressed in different units (raw numbers, percentages, ratios, and so on) must be converted into a common format. For example, before aggregating its data into the Human Development Index,[16] the UNDP converts its data on per capita GDP, life expectancy, and education into variables with a value between 0 and 1. This approach reduces the influence of outliers and the skewness of variable distributions. We chose to preserve this information to a somewhat greater extent, by converting the numbers in our data set into distances from an international average. This distance is expressed as a percentage of the standard deviation of each variable. One consequence of this procedure is that adding a single country to the data set will change the scores of all the other countries. Each country's position on the index, in other words, is entirely relative.

Consistency and Correlation

If the index is to be consistent, each indicator must have the same type of influence on the composite result. For example, if the value of one indicator can range from −1 to +1 and the value of another indicator ranges from 0 to +1, the index is not internally consistent because the latter indicator can never have a negative influence. (In our case the original values were all positive and after conversion, they range from about −2 to +6.) This inconsistency also arises with an indicator that makes a positive contribution when it declines. Finally, different component variables can be substitutes or complements. For

Table A-2. *A Statistical Description of the Indicators*

Enabling factors	Mean	Median	Standard deviation	Skewness	Min-imum	Max-imum
Gross tertiary science enrollment ratio	9.66	9.75	6.38	0.721	0.20	27.40
Per capita GDP	13,193	9,409	9,648	0.470	896	34,142
Resources						
Engineers per million inhabitants	1,333	1,170	1,258	0.871	2	4,960
Institutions per million inhabitants as a percentage of GDP	7.57	2.53	13.80	4.170	0.21	91.75
Production						
Coauthorship index	437	167	652	2.550	2	3,220
Patents per million inhabitants	31.16	1.34	56.36	2.215	0	343.61
S&T journal articles	218.18	92.70	273.19	1.269	0.55	954.92

example, if two indicators are very closely connected, we may actually be double-counting an effect (R close to 1) or inversely measuring it (R close to −1). In this case, the two indicators cancel each other out.

Three tests help to determine the internal consistency of the index. An analysis of the *distribution of values around the mean* shows the degree to which the aggregate is sensitive to variations in each variable. Because the distribution around the mean can be different for each component variable, knowing how these differences will affect the composite index is important. Table A-2 gives summary information about the indicators. The measures of skewness indicate that the distribution of enabling resources (per capita GDP and gross tertiary science enrollment ratio) within our sample is roughly normal. Outliers, however, particularly in the upper ranges, have a more significant impact on the variables representing resources and embedded knowledge. This is particularly true for research institutions, the coauthorship index, and the number of patents.

The second test involves calculating the *correlations between the component variables* to discover substitutes and complementarities. Table A-3 shows that all indicators are positively related. No two indicators cancel each other out. This implies that the selected indicators are complementary. In addition, most of the correlation coefficients are significant at a 1 percent level. The

Table A-3. Correlation Matrix for the S&T Capacity Index

Item	Number of scientists and engineers per million inhabitants	Number of research institutions per million inhabitants	Public funds spent on R&D	Gross science enrollment at the tertiary level	Comparative share of internationally coauthored articles	Patents per million inhabitants	Number of scientific papers published per million inhabitants	GDP in purchasing power parity per capita
Number of scientists and engineers per million inhabitants	1							
Number of research institutions per million inhabitants	0.504(**)	1						
Public funds spent on R&D	0.777(**)	0.502(**)	1					
Gross science enrollment at the tertiary level	0.569(**)	0.332(**)	0.512(**)	1				
Comparative share of internationally coauthored articles	0.562(**)	0.332(**)	0.583(**)	0.295(**)	1			
Patents per million inhabitants	0.734(**)	0.603(**)	0.830(**)	0.340(**)	0.564(**)	1		
Number of scientific papers published per million inhabitants	0.760(**)	0.627(**)	0.859(**)	0.476(**)	0.621(**)	0.764(**)	1	
GDP in purchasing power parity per capita	0.792(**)	0.603(**)	0.795(**)	0.535(**)	0.612(**)	0.742(**)	.855(**)	1

Source: Author's calculations.

**In a Pearson's correlation, correlation is significant at the 0.01 level (two-tailed).

strongest correlations were found between GDP, scientists and engineers, R&D expenditure, the number of articles, and the number of patents. As a consequence, it is possible that some effects are measured twice (for example, patented innovations presented in articles, R&D funds spent on scientists and engineers, or the growth effects of R&D expenditure).

The third test relates to *consistency through time.* The volatility and growth rates of each component can vary considerably. If one component has higher average annual growth or stronger annual fluctuations, over time and at any given moment its impact on the composite index can be more substantial and change more significantly than that of other indicators. Therefore, we tested whether the method of construction takes this possibility into account (these results are not given here).

Based on these three tests, we can conclude that the STCI-02 is internally consistent. Its composite value is not sensitive to the absolute value of its components. All components can range between negative and positive values. They all contribute to the aggregate in the same (positive) manner, as shown by the correlation coefficients. The extent of the contribution of the indicators to the composite index is determined not by their absolute or converted values but by the weights assigned to each component.

WEIGHTING SCHEMES

When the indicators have been properly selected, their values have been converted to a common format, and the internal consistency of the index has been checked, the composite index can be calculated. If the components are expressed in a common unit (such as dollars or number of people), there is no need for weighting. The components, though, are usually not comparable, and even after they have been converted to a common format they cannot simply be added or averaged. To create a composite index, the indicators must be weighted. The following equation shows the method we used to calculate the index and the role that the weights play in this process:

$$ST_i = \frac{\sum_{j=1}^{J} \left(\frac{X_{ij} - \overline{X}_j}{\sigma_j} \right) \cdot W_j}{W} .$$

In this equation

ST_i is the S&T capacity index value for country i

X_{ij} is the value of indicator j for country i

Table A-4. *Weighting Schemes for the S&T Capacity Index*

Weighting scheme number	Enabling factors	Resources	Embedded knowledge
1	1	1	1
2	1	2	1
3	1	3	2
4	1	4	2

$\overline{X}_j$ is the international average of indicator j

σ_j is the standard deviation of indicator j

J is the total number of indicators

W_j is the weight assigned to indicator j

W is the sum of all weights.

Three methods can be used to weight noncomparable indicators:

1. Choosing no weights (all weights are 1)

2. Choosing arbitrary weights after conducting sensitivity analysis to define a range of weighting schemes within which the index is robust

3. Statistically deriving weights through factor analysis.

We implemented the second approach by testing a number of different weighting schemes. Weights were assigned to each of the three S&T capacity domains: enabling factors, resources, and embedded knowledge. The indicators within each domain were not weighted. In each case, we gave resources equal or greater weight than the other two categories of variables because this domain measures capacity most closely. Table A-4 presents the four weighting schemes we tested.

Figure A-1 shows how the different weighting schemes affect the index. The domains of resources and embedded knowledge have a more skewed distribution across countries than that of the enabling factors. When we increase the weight placed on these two domains, then, international differences become more pronounced, with a bias toward the countries that perform better and away from those that lag behind the international mean. Ideally, we would like to minimize this bias. But at the same time we want to give greater weight to the indicators representing resources, since these variables are closest to direct measures of S&T capacity. Therefore, we have chosen to implement the second weighting scheme, which assigns resources twice the weight of the other two domains.

Figure A-1. *Influence of Weighting Scheme on Distributions*

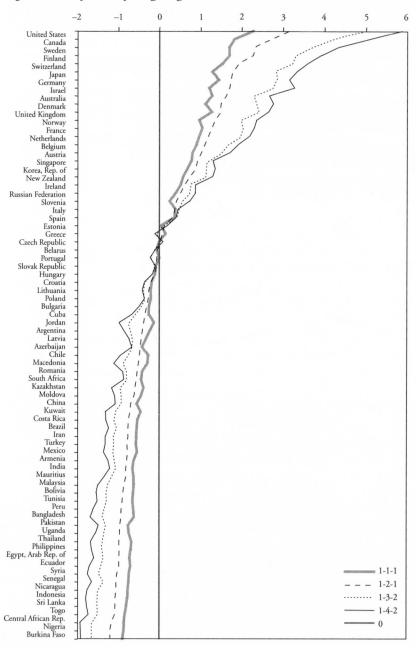

The Science and Technology Capacity Index-2002

The STCI-02 index presents a full analysis of data for 76 countries (refer to table 6-1 for the full ranking of these countries). As described in chapter 6, we divided the countries in our data set into four categories, based on their score on the index:

1. A country is *advanced* if its score is more than one standard deviation above the international average.

2. A country is *proficient* if its score is between the international average and one standard deviation above the international average.

3. A country is *developing* if its score is between the international average and one standard deviation below the international average.

4. A country is *lagging* if its score is more than one standard deviation below the international average.

To refine the classification scheme further, we conducted a statistical cluster analysis.[17] Figure A-2 presents the results. It depicts a set of four advanced clusters on the right, which are closely grouped together, illuminating their interconnectedness. A Canada-U.S. cluster joins with three other clusters— mainly European countries, in addition to Japan and Israel. The other countries are distantly related to the advanced countries, mainly through the United States and Canada.

In further research, the performance of countries by domain will be analyzed. The preconditions, resources, and production of science will be compared within and across countries to achieve an estimate of the productivity of S&T capacity for each country.[18]

Figure A-2.   *Cluster Analysis of the Relationships of Seventy-Six Countries in the Full Index*

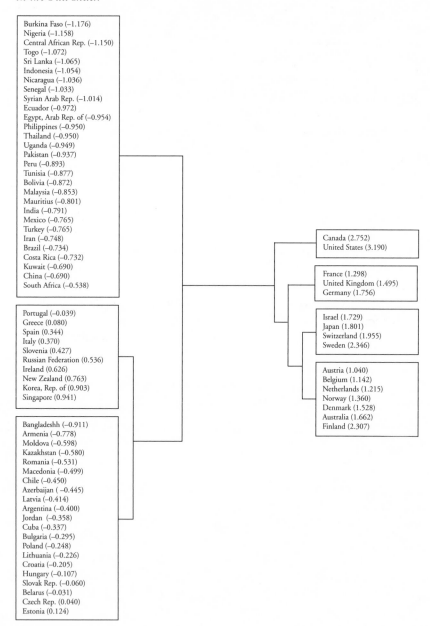

List of Interview Subjects

I extend my deepest gratitude to all the individuals who agreed to be interviewed for this book.

Arthur W. Apter, Department of Mathematics, Baruch College of CUNY, City University of New York, United States

Tran Ngoc Ca, Vietnamese National Institute for Science & Technology Policy and Strategy (NISTPASS), Vietnam

Francisco Chavez-Garcia, Centro de Investigación Sísmica, A.C. Tlalpan, Mexico

Peter L. Collins, Laboratory of Infectious Diseases, National Institute of Allergy and Infectious Diseases, National Institutes of Health, United States

Michael Fehler, Seismic Research Center, Los Alamos National Laboratory, New Mexico, United States

Marco Feroci, Istituto di Astrofisica Spaziale e Fisica Cosmica di Roma, Italy

Joel David Hamkins, City University of New York, College of Staten Island, United States

John Heise, Eramus University National Institute for Space Research (SRON), Netherlands

Robert Hwang, Center for Integrated Nanotechnologies, U.S. Department of Energy's Sandia National Laboratory, United States

Peter Johnston, Evaluation and Monitoring Unit, European Commission, Belgium

Frank E. Karasz, Department of Polymer Science & Engineering, Conte Center for Polymer Research, University of Massachusetts (Amherst), United States

Hans-Dieter Klenk, University of Marburg Virology Institute, Germany

Ulla Lundström, Faculty of Technology and Natural Sciences, Mid-Sweden University, Sweden

Krzysztof Matyjaszewski, Department of Chemistry, Carnegie-Mellon University, United States

Peter Ndemere, Uganda National Science and Technology Council, Uganda

Anand Pillay, Department of Mathematics, University of Illinois at Urbana-Champaign, United States

Luigi Piro, Istituto di Astrofisica Spaziale e Fisica Cosmica di Roma, Italy

Elena Rohzkova, Center for Nanoscale Materials, Argonne National Laboratory, United States

Volker Schonfelder, Max Planck Institute of Extraterrestrial Physics, Germany

Saharon Shelah, Institute of Mathematics, Hebrew University of Jerusalem, Israel, and Rutgers University, Mathematics Department, New Jersey, United States

S. K. Singh, Instituto de Ingeniería, Universidad Nacional Autónoma de México, Mexico

Gerrit ten Brink, Zernike Institute for Advanced Materials, University of Groningen, Netherlands

Wolfgang Wilcke, Institute of Soil Science and Soil Geography, Bayreuth University, Germany

Eckard Wimmer, Department of Microbiology School of Medicine, SUNY at Stony Brook, New York, United States

NOTES

Foreword

1. Anne-Marie Slaughter, "Global Government Network, Global Information Agencies, and Disaggregated Democracy," Public Law Working Paper (Harvard Law School, 2001).

Chapter One

1. Chris Freeman offers an interesting discussion of this question in "Continental, National, and Sub-National Innovation Systems—Complementarity and Economic Growth," *Research Policy* 31 (2002): 191–211.

2. Albert-László Barabási offers an excellent nontechnical explanation of network theory in *Linked: The New Science of Networks* (Cambridge: Perseus Books, 2002).

3. Eric von Hippel, "'Sticky Information' and the Locus of Problem Solving: Implications for Innovation," in *The Dynamic Firm: The Role of Technology, Strategy, Organization, and Regions,* edited by A. Chandler, P. Hagström, and Ö. Sölvell (Oxford University Press, 1999), pp. 60–77.

4. In 1994, two policy books on science and technology pointed out the rise of teaming and integrative research. In *The New Production of Knowledge: The Dynamics of Science and Research in Contemporary Societies* (London: Sage Publications, 1994), Michael Gibbons and others argue that a new form of knowledge production, which they dub Mode Two, began to emerge in the twentieth century. Mode Two is context-driven, problem-focused, and interdisciplinary. It involves multidisciplinary teams brought together for short periods of time to work on specific problems in the real world. John Ziman draws a

similar distinction between academic science and postacademic science in his book, *Prometheus Bound: Science in a Dynamic Steady State* (Cambridge University Press, 1994).

5. In Italian, SAX stands for Satellite per Astronomia X, which translates into English as x-ray astronomy satellite. The satellite was named BeppoSAX in honor of Italian astronomer Giuseppe "Beppo" Occhialini. For more information on BeppoSAX, see the BeppoSAX Mission home page at http://bepposax.gsfc.nasa.gov/bepposax/index.html.

6. For more information on Explorer 11, see NASA's website (http://heasarc.gsfc. nasa.gov/docs/heasarc/missions/explorer11.html).

7. Telephone interview with author, May 27, 2003.

8. One year later, the satellite reentered the earth's atmosphere and fell into the Pacific Ocean.

9. Telephone interview with author, June 6, 2003.

10. Telephone interview with author, June 6, 2003.

11. BeppoSAX Mission home page (http://bepposax.gsfc.nasa.gov/bepposax/index. html)

12. Telephone interview with author, June 6, 2003.

13. Personal interview, June 30, 1999.

14. The ERA is a system of scientific research programs that integrate the EU's scientific resources. Since its creation in 2000, the ERA has concentrated on multinational cooperation in the fields of medical, environmental, industrial, and socioeconomic research. The system can be likened to a research and innovation equivalent of the European common market for goods and services. Its purpose is to increase the competitiveness of European research institutions by bringing them together and encouraging a more inclusive way of work, similar to what already exists among institutions in North America and Japan. Increased mobility of knowledge workers and deepened multilateral cooperation among research institutions in the member states are central goals of the ERA.

15. Caroline S. Wagner and others, *Science and Technology Collaboration: Building Capacity in Developing Countries?* MR-1357.0-WB. (Santa Monica, Calif.: RAND Corporation, 2001).

16. Quoted in Derek de Solla Price, *Science since Babylon* (Yale University Press, 1961), p. 3.

17. L. Biggerio, "Self-Organization Processes in Building Entrepreneurial Networks: A Theoretical and Empirical Investigation," *Human Resources Management* 203 (2001): 209–23.

18. Loet Leydesdorff and Andrea Scharnhorst, *Measuring the Knowledge Base: A Program of Innovation Studies, Report to the Bundesministerium für Bildung und Forschung* (Berlin: Brandenburgische Akademie der Wissenschaft, 2003).

Chapter Two

1. This case study draws on the author's telephone interviews and written exchanges with Wilcke in June and September 2003 while the author was at the University of

Amsterdam. Wilcke was chosen as the subject of this case study because, according to data drawn from the Institute for Scientific Information, he copublished more articles with international partners in 2000 than nearly any other soil scientist.

2. It is common in the United States to call any university teacher "Professor." In European universities, this title is reserved for a few teachers who have shown distinction in teaching, research, and administration.

3. Although three academies founded in Italy predated the Royal Society of London, they were not in continuous existence. The Accademia Secretorum Naturae was formed in Naples in 1560; the Accademia del Cimento was founded in Florence in 1651; and the Accademia dei Lincei existed in Rome from 1603 to 1630. Louis XIV chartered the Paris Academie in 1666. In 1661, Charles II chartered the Royal Society of London, the group emerging from the original invisible college.

4. By 1830, according to Derek de Solla Price, so many journals had been created that a new invention was required to handle the new knowledge: researchers were forced to invent the "abstract journal" to summarize subfields of knowledge. See Derek de Solla Price, *Little Science, Big Science* (Columbia University Press, 1963).

5. This process of exponential increase in published scientific literature has been noted at several points in the development of science, notably by F. Galton, *Heredity Genius* (London: MacMillan, 1869); A. Lotka, "The Frequency Distribution of Scientific Productivity," *Journal of the Washington Academy of Sciences* 16 (1926); and Price, *Little Science, Big Science.*

6. E. N. da C. Andrade, *A Brief History of the Royal Society, 1660–1960* (London: The Royal Society, 1960). In the mid-seventeenth century in England, Andrade wrote, "it was widely held that the writings of the great classical philosophers were the one fountain of wisdom, whence all knowledge flowed. In particular, the writings of Aristotle were the great authority on all scientific matters, such as the laws of terrestrial mechanics, the motions of the heavens, and the nature of light and color. . . . In fact, as late as 1692 Sir William Temple, in his essay Upon the Ancient and Modern Learning, set out to prove that all wisdom lay with the ancients" (Andrade, *Brief History,* p. 6).

7. Andrade, *Brief History,* p. 4.

8. Andrade, *Brief History,* p. 25. This expression, according to the Royal Society, continues to show "its enduring commitment to empirical evidence as the basis of knowledge about the natural world." I. Masson has translated this phrase as follows: "Not under bond to abide by a master's authority," in Horace's *Epistles,* no. I, cited in Dorothy Stimson, *Scientists and Amateurs: The History of the Royal Society* (New York: H. Shuman, 1949).

9. T. Sprat, *A History of the Royal Society* (London: Angel & Crown 1722, corrected version), p. 56.

10. W. C. Dampier, *A History of Science and Its Relation with Philosophy and Religion* (Cambridge University Press, 1929), p. 149.

11. Dampier, *A History of Science,* p. 149.

12. Butterfield, *Origins of Modern Science: 1300–1700* (London: Free Press, 1957).

13. Stimson, *Scientists and Amateurs,* p. 86.

14. See note 3.

15. Sprat, *History of the Royal Society,* p. 61.

16. Dorothy Stimson, *Scientists and Amateurs, A History of the Royal Society* (New York: Greenwood Press, second printing, 1968).

17. Rudolph Stichweh, "Science in the System of World Society," *Social Science Information* 35 (1996): 327–40.

18. Donald deB. Beaver and R. Rosen, "Studies in Scientific Collaboration, Part I. The Professional Origins of Scientific Coauthorship," *Scientometrics* 1 (1978): 65–84.

19. Ibid., p. 66.

20. Ibid., p. 67.

21. Before the passage of the Statute of Monopolies in 1623, the Crown had frequently resorted to the "sale of letters patent"—in which monopolies were granted for the sale of goods or services—to raise funds. Through the statute, Parliament sought to curb this practice.

22. J. A. Schumpeter, *Change and the Entrepreneur* (Harvard University Press, 1949).

23. Beaver and Rosen, "Studies in Scientific Collaboration."

24. Price, *Little Science, Big Science,* pp. 33–34.

25. These numbers are derived from the author's reviews of those portions of the Budget of the United States relevant to science and technology.

26. OECD, *OECD Science, Technology and Industry Scoreboard 2005* (Paris: OECD Publishing, 2005).

27. National Science Board, *Science & Engineering Indicators 2006* (Arlington, Va.: National Science Foundation [NSF], 2006).

28. This estimate is based on an NSF study in which the rate of research spending dedicated to infrastructure is estimated. The report is described in more detail in chapter 6 of this volume. National Science Board, *Science and Engineering Infrastructure for the 21st Century: The Role of the National Science Foundation* (Arlington, Va.: NSF, 2003).

29. Price, *Little Science, Big Science,* p. 17.

30. Vannevar Bush, *Science, the Endless Frontier* (Arlington, Va.: NSF, 1989), available at (www.nsf.gov/od/lpa/nsf50/vbush1945.htm).

31. Robert Schmookler, *Invention and Economic Growth* (Harvard University Press, 1966), p. 177. Schmookler showed that science and technology can be considered endogenous to economic activity, which represented an important departure from neoclassical economic thought. Although Adam Smith and Karl Marx both note the importance of science and technology to economic growth, neoclassical economists considered them to be exogenous factors. In 1956, Robert Solow developed a groundbreaking neoclassical growth model that found that land, labor, and capital could account for only part of economic growth. The "residual" appeared to be attributable to knowledge embedded in technology. Even so, others have observed that no explicit model is yet capable of determining causal relationships between science, technology, and economic growth.

32. In the public realm, these types of institutions included those dedicated to funding basic science, such as the NSF in the United States. Other agencies, though mission-

oriented, had a significant basic and applied R&D budget. The U.S. departments of defense and energy are examples of these agencies.

33. The governments of a number of nations created extension services and local centers to offer technical support in agriculture and engineering.

34. Loet Leydesdorff and Henry Etzkowitz, "A Triple Helix of University-Industry-Government Relations: Mode 2 and the Globalization of National Systems of Innovation," in *Science under Pressure* (Aarhus: Danish Institute for Studies in Research and Research Policy, 2001).

35. Paul Bairoch and Maurice Levy-Leboyer, *Discrepancies in Economic Growth since the Industrial Revolution* (New York: St. Martin's Press, 1981).

36. World Bank, *World Development Report: The Challenge of Development* (Oxford University Press, 1991).

37. David Smith and J. Sylvan Katz, *Collaborative Approaches to Research, HEFCE Fund Review of Research Policy and Funding, Final Report* (University of Sussex, April 2000).

38. OTA, *International Partnerships in Large Science Projects*, OTA-BP-ETI-150 (Washington: U.S. Government Printing Office, July 1995).

39. See Peter Galison and Bruce Hevly, eds., *Big Science: The Growth of Large-Scale Research* (Stanford University Press, 1992), and Karin D. Knorr-Cetina, *Epistemic Cultures: How the Sciences Make Knowledge* (Harvard University Press, 1999).

40. See www.gbif.org.

41. This point has been made in a number of publications, including Michael Gibbons and others, *The New Production of Knowledge: The Dynamics of Science and Research in Contemporary Societies* (London: Sage Publications, 1994), and John Ziman, *Prometheus Bound: Science in a Dynamic Steady State* (Cambridge University Press, 1994).

42. Nicholas Vonortas, *Cooperation in Research and Development* (Boston: Kluwer, 1997).

43. Based on the author's review of R&D budgets of scientifically advanced countries.

44. The Human Frontiers Science Program (see www.hfsp.org/) provides research grants for teams of scientists from different countries who wish to combine their expertise to approach questions that could not be answered by individual laboratories. Emphasis is placed on novel collaborations that bring together scientists from different disciplines (for example, from chemistry, physics, computer science, and engineering) to focus on problems in the life sciences. Funding is furnished primarily by the government of Japan, with support from other G-7 nations, as well as by Australia, India, the Republic of Korea, Switzerland, New Zealand, and the non-G-7 members of the European Union, who are represented by the European Commission.

Chapter Three

1. Robert Axelrod, *The Complexity of Cooperation: Agent-Based Models of Competition and Collaboration* (Princeton University Press, 1997), p. 4.

2. K. Boyack, R. Klavans, and K. Börner, "Mapping the Backbone of Science," *Scientometrics* 64, no. 3 (August 2005): 351–74.

3. Many articles address this phenomenon; in particular, see L. Leydesdorff, "The Delineation of Nanoscience and Nanotechnology in Terms of Journals and Patents: A Most Recent Update," *Scientometrics* (2009, forthcoming).

4. As an example, see Caroline S. Wagner and others, *Evaluation of NETworks of Collaboration among Participants in IST Research and Their Evolution to Collaborations in the European Research Area (ERA)*, Monograph 254-EC (Leiden, Netherlands: RAND Europe, 2005).

5. C. Lee Giles, Isaac G. Councill, and J. N. Gray, "Who Gets Acknowledged? Measuring Scientific Contributions through Automatic Acknowledgement Indexing," *Proceedings of the National Academy of Science* 101, no. 51 (2004): 17599–604.

6. It does not appear that each actor must give and take in equal measure—some may give more and some may take more. The question seems to be whether the network as a whole balances the give-and-take communications.

7. Complementary capabilities are forged when researchers from two fields collaborate to improve research outcomes. The two fields, however, maintain the integrity of their boundaries and the collaborative research does not contribute to defining a new field. When researchers from two fields come together to create a new field of study, that research is said to be "integrative."

8. Francis Fukuyama, *Trust: The Social Virtues and the Creation of Prosperity* (New York: Free Press, 1995).

9. Fukuyama, *Trust,* p. 26.

10. Political scientist Robert D. Putnam was largely responsible for the resurgence of interest in social capital that began in the mid-1990s. See Robert D. Putnam, Robert Leonardi, and Raffaella Y. Nanetti, *Making Democracy Work: Civic Traditions in Modern Italy* (Princeton University Press, 1994); and Robert D. Putnam, *Bowling Alone: The Collapse and Revival of American Community* (New York: Simon & Schuster, 2000).

11. The benefits of social capital were enumerated by James Coleman, "Social Capital in the Creation of Human Capital," *American Journal of Sociology* 94 *(Supplemental: Organizations and Institutions: Sociological & Economic Approaches to the Analysis of Social Structure,* 1988): s95–s120.

12. Telephone interview with the author, September 17, 2003.

13. Thomas Kuhn, *The Structure of Scientific Revolutions* (University of Chicago Press, 1962).

14. The probability that any node, *k,* is connected to any other node is proportional to $1/kn$. According to Albert-László Barabási and Eric Bonabeau, the value of *n* tends to fall between 2 and 3. Barabási and Bonabeau, "Scale-Free Networks," *Scientific American* 288, no. 5 (May 2003): 60–69.

15. Ibid.

16. According to Mark Newman, "The power law is a distinctive experimental signature seen in a wide variety of complex systems. In economics it goes by the name 'fat tails,' in physics it is referred [to] as 'critical fluctuations,' in computer science and biology it is 'the edge of chaos,' and in demographics and linguistics it is called 'Zipf's law.'" Newman points to the possibility that "several other features of many complex systems including

robustness to perturbations and sensitivity to structural flaws, may be the result of the design or evolution of systems for optimal behaviour." M. E. J. Newman, "The Power of Design," *Nature* 405 (May 25, 2000): 412–13. See also note 22.

17. Barabási and Bonabeau, "Scale-Free Networks."

18. Albert-László Barabási and Reka Albert, "Emergence of Scaling in Random Networks," *Science* 286 (October 1999): 509–15.

19. Michaelis Faloutsos, Petros Faloutsos, and Christos Faloutsos, "On Power Relationships of the Internet Topology," in *Proceedings of the ACM SIGCOMM '99 Conference on Applications, Technologies, Architectures, and Protocols for Computer Communication, August 30–September 3, 1999* (Cambridge, Mass.: Association for Computing Machinery [ACM], 1999), pp. 252–62.

20. Herbert Simon, "On a Class of Skewed Distribution Functions," *Biometrika* 42 (1955): 425–40.

21. A. Lotka, "The Frequency Distribution of Scientific Productivity," *Journal of the Washington Academy of Sciences* 16 (1926): 317.

22. Other early investigations into power-law distributions were done by Simon, "On a Class of Skewed Distribution Functions," and by applying Zipf's law, presented in G. K. Zipf, *Human Behavior and the Principle of Least Effort.* (New York: Hafner, 1949).

23. Mark Newman, "Who Is the Best Connected Scientist? A Study of Scientist Coauthorship Networks," *Physics Review E* 64 016132 (2001): 7 pages.

24. Barabási and Bonabeau, "Scale-Free Networks."

25. Barabási and Albert, "Emergence of Scaling in Random Networks." Derek de Solla Price called this process "cumulative advantage" in *Little Science, Big Science,* (Columbia University Press, 1963), p. 43. Sociologist Robert Merton used the term *Matthew effect,* recalling the biblical observation that the rich get richer, to refer to the related phenomenon in which well-known authors tend to get more credit than their coauthors for joint work. Merton, "The Matthew Effect in Science," *Science* 159, no. 3810 (1968): pp. 56–63.

26. In a study of evolving networks, several research groups have shown that highly connected nodes increase their connectivity faster than their less connected peers. See Barabási and others, "Evolution of the Social Network of Scientific Collaborations," *Physica A: Statistical Mechanics and Its Applications* 311, nos. 3 and 4 (August 15, 2002): 590–614.

27. Mark Granovetter, "The Strength of Weak Ties," *American Journal of Sociology* 78, no. 6 (May 1973): 1360–80, and "The Strength of Weak Ties: Network Theory Revisited," *Sociological Theory* 1 (1983): 201–33.

28. Peter Csermely, *Weak Links: Stabilizers of Complex Systems from Proteins to Social Networks* (Berlin: Springer, 2006).

29. Granovetter, "The Strength of Weak Ties."

30. Personal interview with author, November 20, 2007.

31. Stanley Milgram popularized this term with a series of experiments devised to examine the average path length within social networks in the United States. His groundbreaking research revealed that human society is a small-world type of network characterized by shorter-than-expected path lengths. The experiments are often associated with the term

six degrees of separation, although Milgram did not use this term himself. Milgram, "The Small World Problem," *Psychology Today* 1 (May 1967): 60–67.

32. The term was popularized by John Guare's play "Six Degrees of Separation," which was later made into a movie.

33. Craig Fass, Brian Turtle, and Mike Ginelli, *Six Degrees of Kevin Bacon* (New York: Plume, 1996).

34. Small worlds are identified as networks in which local clustering is high and global path lengths across the network are small. Duncan Watts and Steven Strogatz found that small worlds can be produced by adding even a few extra links to an ordered network. Watts and Strogatz, "Collectives Dynamics of 'Small-World' Networks," *Nature* 393 (1998): 440–42.

35. Newman, "Who Is the Best Connected Scientist?"

36. Redundancy is also related to density, a network measure calculated by dividing the total number of links within a network by the number of potential links. Dense networks have many opportunities for exchange and many possible pathways from any one node to any other node in the network. Sparse or tightly clustered networks have fewer opportunities for information to move between the nodes in a network.

37. Jean Piaget, "John Amos Comenius (1592–1670)," *Prospects* (UNESCO, International Bureau of Education) XXIII, no. 1/2 (1993): 173–96. See www.ibe.unesco.org/publications/ThinkersPdf/comeniuse.PDF.

38. *The Hartlib Papers,* 2nd ed., held at Sheffield University Library, available at (www.shef.ac.uk/library/special/hartlib.html).

39. Comenius's original document was entitled *Pansophiae Prodromus.* When it was published by Oxford University, it was retitled *Conatum Comenianorum Praeludia.*

40. Comenius believed the invitation to England came from Parliament, but that was not the case, according to Dorothy Stimson, "Comenius and the Invisible College," *Isis* 23, no. 2 (September 1935): 373–88.

41. D. Stimson, *Scientists and Amateurs, A History of the Royal Society.* (New York: Greenwood Press, second printing, 1968).

42. This bit of history—the role of Comenius in suggesting the formation of an invisible college and the role of the German nationals in organizing the first meetings of the Royal Society members—is not often cited in the official histories of the Royal Society, such as E. N. da C. Andrade's *A Brief History of the Royal Society, 1660–1960* (London: The Royal Society, 1960). In these histories, the British intellectuals play a more prominent—and sometimes the only—role in setting down the foundation of science. Stimson details the processes by which these individuals met each other in *Scientists and Amateurs* and "Comenius and the Invisible College." With regard to who initiated the idea of such an invisible college, Comenius is presumed to have introduced it at a meeting during his stay in England between 1641 and 1642, but Boyle was the first to use the idea in print, in a letter to his tutor in 1645.

43. D. Harkness, "Entering the Labyrinth: Exploring Scientific Culture in Early Modern England," *Journal of British Studies* 37, no. 4 (October 1998): 446–50.

44. Kuhn, *The Structure of Scientific Revolutions.*

45. Ibid.

46. Ibid.

47. Telephone interview between Susan A. Mohrman, Brookhaven National Laboratory, and Bob Hwang, November 18, 2005.

48. New ideas do not emerge randomly. Previous investments, the location and sophistication of laboratories, and the ideas of talented people all contribute to what network theorists would call "path dependency"—the effect of past conditions on future possibilities.

Chapter Four

1. H. H. Hess, "History of Ocean Basins," in *Petrologic Studies: A Volume in Honor of A. F. Buddington,* edited by A. E. J. Engel, H. L. James, and B. F. Leonard (New York: Geological Society of America, 1962), pp. 599–620, and R. S. Deitz, "Continent and Ocean Basin Evolution by Spreading of the Sea Floor," *Nature* 190 (1961): 854–57. It is not uncommon in science for a breakthrough theory to be proposed independently by two people, which appears to have taken place with the theories underlying calculus, natural selection, and the double helix.

2. H. Sato and M. Fehler, *Seismic Wave Propagation and Scattering in the Heterogeneous Earth* (Washington: American Institute of Physics Press, 1998).

3. Telephone interview with author, October 18, 2005.

4. C. S. Wagner and L. Leydesdorff, "Measuring the Globalization of Knowledge Networks," in *Blue Sky II Forum 2006, Organization for Economic Cooperation and Development (OECD) Conference Proceedings* (2006). See (www.oecd.org/document/24/0,3343, en_2649_201185_37075032_1_1_1_1,00.html).

5. C. S. Wagner and L. Leydesdorff, "Seismology as a Case Study of Distributed Collaboration in Science," *Scientometrics* 58, no. 1 (2003): 91–114.

6. C. S. Wagner and L. Leydesdorff, "Mapping Global Science Using International Coauthorships: A Comparison of 1990 and 2000," *International Journal of Technology and Globalization* 3 (2005): 185–92.

7. O. Persson and others, "Inflationary Bibliometric Values: The Role of Scientific Collaboration and the Need for Relative Indicators in Evaluative Studies," *Scientometrics* 60, no. 3 (August 2004): 421–32.

8. W. Glänzel and others, "A Bibliometric Analysis of International Scientific Cooperation of the European Union (1985–1995)," *Scientometrics* 45, no. 2, (1999), and F. Narin, "Globalisation of Research, Scholarly Information and Patents—Ten Year Trends," in *Proceedings of the North American Serials Interest Group (NASIF) 6th Annual Conference, The Serials Librarian* 21 (1991): 2–3.

9. It is important to note that these measures reflect only formal communications. Recall that formal communications sit atop a great deal of more informal interaction, suggesting that the network of science is even denser than publication-based graphs show.

10. These fields could be considered subfields of even larger fields. Mathematical logic is a subfield of mathematics, virology of biology, polymers of chemistry, soil science of

agronomy, and so on. The subfield level is much more specific and can be analyzed with more precision than data handled at the field level.

11. Loet Leydesdorff and I found that the global rate of growth of international collaboration in science was 15.6 percent between 1990 and 2000, as reported in "Mapping Global Science."

12. Because seismology did not exist as a separate field in 1990, I treat it and geophysics as one scientific field for the purposes of this chapter. A separate study, however, reveals that seismology emerged from within geophysics during the study period. See Wagner and Leydesdorff, "Seismology as a Case Study."

13. This research is presented in several published papers: Wagner and Leydesdorff, "Network Structure, Self-Organisation and International Collaboration in Science," *Research Policy* 34, no. 10 (2005): 1608–18; Wagner, "Six Case Studies of International Collaboration in Science," *Scientometrics* 62, no. 1 (2005): 3–26; Wagner and Leydesdorff, "Mapping Global Science"; and Wagner and Leydesdorff, "Seismology as a Case Study."

14. M. Gibbons and others, *The New Production of Knowledge: The Dynamics of Science and Research in Contemporary Societies* (London: Sage, 1994).

15. R. K. Merton, *Social Theory and Social Structure* (Columbia University Press, 1957), and R. Whitley, *The Intellectual and Social Organisation of the Sciences* (Oxford University Press, 1984).

16. John Holland, *Hidden Order: How Adaptation Builds Complexity* (Reading, Mass.: Helix Books, 1995).

17. John Holland, *Emergence: From Chaos to Order* (New York: Addison-Wesley, 1998).

18. Telephone interview with author, March 18, 2003.

19. Ibid.

20. F. Sagasti, *The Sisyphus Challenge: Knowledge, Innovation and the Human Condition in the 21st Century* (Lima, Peru: FORO Nacional, 2003).

21. Personal interview with author, December 9, 2007.

22. Telephone interview with author, November 21, 2002.

23. P. Hoffman, *The Man Who Loved Only Numbers: The Story of Paul Erdos and the Search for Mathematical Truth* (New York: Hyperion, 1998).

24. Erdos' productivity inspired some of his colleagues to attempt to categorize all mathematicians according to their closeness to Erdos. Someone who coauthored an article with Erdos is considered to have an Erdos number of one. Someone who coauthored an article with an Erdos coauthor has an Erdos number of two, and so on. To date, the Erdos Number Project has identified 268,000 people with a finite Erdos number. (Individuals who cannot be connected to Erdos are considered to have an infinite Erdos number.) The maximum identified Erdos number is 13, and the mean is 4.65, indicating that the average person within this network is fewer than five steps away from Erdos—another illustration of the small-world phenomenon. These data are provided by the Erdos Number Project at www.oakland.edu/enp/trivia.html.

25. G. Laudel, "Collaboration, Creativity and Rewards: Why and How Scientists Collaborate," *International Journal of Technology Management* 22 (2001): 762–81.

26. A. Saxenian, *Regional Advantage: Culture and Competition in Silicon Valley and Route 128* (Harvard University Press, 1996).

27. National Science Board, *Science and Engineering Indicators 2000* (Arlington, Va.: NSF, 2001).

28. C. S. Wagner and others, *Science & Technology Cooperation: Building Capacity in Developing Countries,* Monograph-1357-WB (Santa Monica, Calif.: RAND Corporation, 2001).

29. Saxenian, *Regional Advantage.*

30. National Science Board, *Science and Engineering Indicators 2002* (Arlington, Va.: NSF, 2003).

31. UNESCO, *Global Education Digest, 2006: Comparing Education Statistics across the World* (Quebec: UNESCO Institute for Statistics, 2006).

32. R. Bhandari, "Institute for International Education Project Atlas," presented to the Sigma Xi Conference on Science, Technology and the Future of the Workforce, Washington, D.C., September 20, 2006.

Chapter Five

1. There is a great deal of literature on regional economic development and its relationship to science and technology. The question of the relationship between the two remains a point of continuous debate in science and technology studies. One of the most interesting of these is M. Zitt and others, "Potential Science-Technology Spillovers in Regions: An Insight on Geographic Co-Location of Knowledge Activities in the EU," *Scientometrics* 57 (2003): 295–320. This article shows an uneven relationship between science and technology intensity of regions and the occurrence of economic growth. Nathan Rosenberg treats this subject by discussing the exogeneity of science in *Inside the Black Box: Technology and Economics* (Cambridge University Press, 1982). See also AnnaLee Saxenian, *Regional Advantage: Culture and Competition in Silicon Valley and Route 128* (Harvard University Press, 1994).

2. M. R. Pinto and others, "Light-Emitting Copolymers of Cyano-Containing PPV-Based Chromophores and a Flexible Spacer," *Polymer* 41, no. 7 (March 2000): 2603–11.

3. Telephone interview with author, July 10, 2003.

4. National Science Board, *Science and Engineering Indicators 2008* (Arlington, Va.: National Science Foundation, 2008).

5. Ibid.

6. W. Brian Arthur, *Increasing Returns and Path Dependence in the Economy* (University of Michigan Press, 1994).

7. One study that was very influential on my thinking is C. Freeman, "Continental, Sub-Continental, and National Innovation Systems—Complementarity and Economic Growth," *Research Policy* 31 (2002): 191–211. See also J. Mokyr, *The Gifts of Athena* (Princeton University Press, 2002).

8. For an overview of this literature, see Gordon L. Clark, Meric S. Gertler, and Maryann P. Feldman, eds., *The Oxford Handbook of Economic Geography* (Oxford University Press, 2003).

9. The use of online resources such as real-time design and testing is growing into a practice some call "eScience." The eScience Institute has resources about this at www.nesc. ac.uk/esi/.

10. For more information on this project, see setiathome.berkeley.edu.

11. A. Ferreira and C. Mavroidis, "Virtual Reality and Haptics for Nanorobotics," *IEEE Robotics and Automation Magazine* (September 2006): 78–83.

12. J. Mendler, D. Simon, and P. Broome, "Virtual Development and Virtual Geographies: Using the Internet to Teach Interactive Distance Courses in the Global South," *Journal of Geography in Higher Education* 26, no. 3 (November 1, 2002): 313–25.

13. In the many interviews I conducted with scientists who were working at the international level on distributed and coordinated research projects, I found that the vast majority of these projects began with or early on had a face-to-face meeting.

14. John Rawls, *A Theory of Justice* (Cambridge: Belknap Press, 1971), p. 17.

15. Rawls, *A Theory of Justice*, p. 150.

16. Francis Fukuyama, *Trust: The Social Virtues and the Creation of Prosperity* (New York: Free Press, 1995).

17. Personal interview with author, Kampala, Uganda, July 5, 2007.

18. Judi Wakhungu, "Public Participation in Science and Technology Policymaking: Experiences from Africa," 2004; see practicalaction.org/?id=publicgood_wakhungu.

19. Personal interview with author, Kampala, Uganda, July 5, 2007.

Chapter Six

1. C. Freeman, "Continental, Sub-Continental, and National Innovation Systems—Complementarity and Economic Growth," *Research Policy* 31 (2002), cites the 1991 *World Development Report* as showing an increasing disparity of growth among different parts of the world. The World Development Report is issued each year by the World Bank; the most recent of these reports can be found at the following link (http://econ.worldbank.org/wdr/).

2. Jeffrey Sachs, *The End of Poverty* (New York: Penguin Press, 2004), p. 70.

3. See (www.naro.go.ug).

4. Informal workshop, Los Alamos National Laboratory, New Mexico, Spring 2006.

5. As quoted in Stephen Inwood, *The Forgotten Genius: The Biography of Robert Hooke 1635–1703* (London: MacAdam/Cage, 2005).

6. This work was done in collaboration with Edwin Horlings of the Rathenau Instituut (The Hague, Netherlands) and Arindam Dutta, RAND Corporation (Santa Monica, Calif.). It updates the RAND Corporation's 2000 index of science and technology capacity (C. S. Wagner and others, *Science & Technology Cooperation: Building Capacity in Developing Countries*, Monograph 1357-WB [RAND Corporation, 2001]).

7. Ibid.

8. Bush, *Science, the Endless Frontier,* A Report to the President by Vannevar Bush, Director of the Office of Scientific Research and Development, July 1945 (Washington: United States Government Printing Office, 1945).

9. Telephone interview with the author, February 23, 2007.

10. According to UNESCO, the primary net enrollment rate has increased from 86 percent in 1990 to 91 percent in 2003, the dropout rate has declined from 12 percent to about 3 percent, the repetition rate has dropped from 9 percent to less than 5 percent, and the completion rate has increased from 47 percent to more than 75 percent. But a recent national school survey, which suggests a lower level of progress, is used as the baseline for Ca's work. In addition, Vietnam has also significantly expanded lower secondary education opportunities. The transition rate from primary to lower secondary has increased from 78 percent to 88 percent, allowing a majority of young Vietnamese to gain access to nine years of basic education. The education sector's share of total public expenditure in 2003 was 17 percent.

11. Telephone interview with the author, February 23, 2007.

12. The 7th International Symposium on General Convexity/Monotonicity was held in Hanoi in August 2002.

13. T. Kuwahara, "Technology Forecasting Activities in Japan—Hindsight on 30 Years of Delphi Expert Surveys," *Technological Forecasting and Social Change* 60, no. 1 (January 1999): 5-14(10), Elsevier.

14. The World Bank, *World Development Indicators 2002* (Washington: World Bank, 2002).

15. National Science Board, *Science and Engineering Infrastructure for the 21st Century: The Role of the National Science Foundation* (Arlington, Va.: NSF, 2003).

16. Ibid.

17. Ibid.

18. Ibid.

19. Ibid.

20. A partial exception to this rule is the proportion of R&D spending that is typically allocated to cover overhead costs, which include the maintenance of capital equipment and buildings—defined as an operating cost—but not the construction of new facilities.

21. This is true across the entire industrialized base of an economy, and not just for those sectors related to science, but this discussion focuses on the parts of these functions that relate to science.

22. See the NIST Quantum Physics Division Physics Laboratory website at (physics.nist.gov/Divisions/Div848/div848.html).

23. According to the Organization for Economic Cooperation and Development (OECD), the 2002 budget of the Japanese Industrial Standards Center for safety, metrology, and standards was approximately US$110.3 million. The EU spends more than €83 billion per year, representing nearly 1 percent of the EU's GDP, on measurement and standardization, generating an estimated €3 of benefit for every €1 spent on measurement

activity and the United States spends an equivalent amount. In the developed countries, industry bears a hefty share of the costs and fees of standardization and metrology. See G. Williams, *The Assessment of the Economic Role of Measurements and Testing in Modern Society* (Oxford: European Commission, 2002).

24. C. S. Wagner and M. Reed, *The Pillars of Progress, Infrastructure for Science* (Vienna: UNESCO, 2004).

25. According to Japanese government websites, the Kohsetsushi engineering centers received approximately US$500 million ($400,000 per 100,000 inhabitants) in cumulative funding during fiscal year 1988.

26. In Germany, technology transfer centers had a budget of approximately US$95 million ($116,000 per 100,000 inhabitants) in 1995.

27. C. S. Wagner and A. Yezril, *Global Science and Technology Information: A New Spin on Access,* Monograph Report-1079-OSTP (Santa Monica, Calif.: RAND Corporation, 1999). See rand.org/publications/MR/MR1079/.

28. These are detailed in D. Nelkin, *Science as Intellectual Property: Who Controls Research?* (New York: Macmillan Press, 1984).

29. Aaron S. Kesselheim and Jerry Avorn, "University-Based Science and Biotechnology Products, Defining the Boundaries of Intellectual Property," *JAMA* 293 (2005): 850–54.

30. David C. Mowery and others, "The Growth of Patenting and Licensing by U.S. Universities: An Assessment of the Effects of the Bayh–Dole Act of 1980," *Research Policy* 30, no. 1 (January 2001): 99–119. Other analysts find that changes in the intellectual property rights regime have not substantially altered the norms that guide scientific research. See Fiona Murray and Scott Stern, "Do Formal Intellectual Property Rights Hinder the Free Flow of Scientific Knowledge? An Empirical Test of the Anti-Commons Hypothesis," National Bureau of Economic Research (NBER) Working Paper No. 11465 (July 2005). See nber.org/papers/w11465.

Chapter Seven

1. With the addition of the eastern European scientists, it is possible to see a single world system of science, as described in chapter 4.

2. Telephone interview with author, April 3, 2003.

3. A considerable body of literature exists around the topic of national innovation systems. See B. Å. Lundvall, *National Innovation Systems: Towards a Theory of Innovation and Interactive Learning* (London: Pinter, 1992).

4. R. R. Nelson and H. Pack, *The Asian Miracle and Modern Growth Theory* (Washington, D.C.: World Bank, 1997).

5. F. Teng-Zeng and J. Mouton. *Innovation Systems within the Context of Socioeconomic Development and Transformation in Africa* (Stellenbosch, South Africa: Stellenbosch University Centre for Research on Science & Technology, January 2006).

6. Private businesses fund some basic research. According to the National Science Board, universities and colleges have historically been the largest performers of basic research in the United States, and in recent years they have accounted for more than half of the nation's basic research (55 percent in 2004). Most basic research is federally funded. The long-term trend in the United States, however, has been to reduce the percentage share of government spending on research and development (R&D) and to increase the percentage share of private spending, even as both parties have increased spending overall (National Science Board, *Science and Engineering Indicators 2004* [Arlington, Va.: National Science Foundation, 2004]).

7. Paul A. David, David Mowery, and W. Edward Steinmueller, "Analyzing the Economic Payoffs from Basic Research," *Economics of Innovation and New Technology* 2, no. 1 (1992): 73–90.

8. As these examples indicate, identifying a pure public good can be difficult. Most public goods in practice display some degree of rivalness or excludability.

9. I. Kaul, I. Grunberg, and M. Stern, *Global Public Goods: International Cooperation in the 21st Century* (Oxford University Press, 1999).

10. Social rates of return are difficult to measure. But see E. Mansfield, "Social and Private Rates of Return from Industrial Innovation," *Quarterly Journal of Economics* 91 (1977): 221–40, and S. W. Popper, *Economic Approaches to Measuring the Performance and Benefits of Fundamental Science* (Santa Monica, Calif.: RAND Corporation, 1995).

11. Judi Wakhungu, "Public Participation in Science and Technology Policymaking: Experiences from Africa," 2004 (practicalaction.org/?id=publicgood_wakhungu).

12. For a complete list of the goals and more about the project, see un.org/millennium goals/ and the associated report by the U.N. Millennium Development Task Force on Science, Technology, and Innovation, which can be found at (www.unmillenniumproject. org/documents/tf10apr18.pdf).

13. See (arxiv.org/).

14. Caroline S. Wagner and others, *Science and Technology Collaboration: Building Capacity in Developing Countries?* MR-1357.0-WB (Santa Monica, Calif.: RAND Corporation, 2001).

Appendix A

1. P. A. David and D. Foray, "An Introduction to the Economy of the Knowledge Society," *International Social Science Journal* 171 (March 2002). Also see chapters on the knowledge base by D. Foray and B.-A. Lundvall and by M. Abramowitz and P. A. David in *Employment and Growth in the Knowledge-Based Economy* (Paris: Organization for Economic Cooperation and Development [OECD], 1996). A large body of literature addresses the question of S&T's contribution to economic growth and knowledge creation. Among these are M. Gibbons and others, *The New Production of Knowledge: The Dynamics of Science and Research in Contemporary Societies* (London: Sage Publications,

1994) and Richard R. Nelson, *The Sources of Economic Growth* (Harvard University Press, 2000).

2. This work updates the RAND Corporation's 2000 index of S&T capacity with more recent data and expands the number of countries that have full data coverage from 66 to 76.1. For the 2000 index, see C. S. Wagner and others, *Science & Technology Cooperation: Building Capacity in Developing Countries,* Monograph 1357-WB (Santa Monica, Calif.: RAND Corporation, 2000).

3. United Nations Development Programme (UNDP), *Human Development Report 2001: Making Technologies Work for Human Development* (Oxford University Press, 2001), and D. Archibugi and A. Coco, "A New Indicator of Technological Capabilities for Developed and Developing Countries (ArCo)," *World Development* 32, no. 4 (April 2004): 629–54.

4. L. Leydesdorff, "The Knowledge-Based Economy and the Triple Helix Model," chapter 2 in *Reading the Dynamics of a Knowledge Economy,* edited by Wilfred Dolfsma and Luc Soete (Cheltenham: Edward Elgar, 2006): 42–76.

5. Braczyk, H.-J., P. Cooke, and M. Heindenreich, eds., *Regional Innovation Systems* (London: University College Press, 1998).

6. C. S. Wagner and L. Leydesdorff, "Mapping the Global Network of Science: A Comparison of 1990 and 2000," *International Journal of Technology and Globalisation* 1, no. 2 (2005): 185–208.

7. UNDP, *Human Development Report 2002, Deepening Democracy in a Fragmented World* (New York: Oxford University Press, 2002), and *CIA World Factbook 2001* (Washington, D.C.: Central Intelligence Agency, 2001).

8. UNDP, *Human Development Report 2002.*

9. Ibid.

10. World Bank, *World Development Indicators 2002* (Washington: World Bank).

11. UNDP, *Human Development Report 2002.*

12. See www.uspto.gov.

13. National Science Board, *Science and Engineering Indicators 2000* (Arlington, Va.: National Science Foundation, 2004).

14. This measure of shares of international collaboration was created specifically for this work using data from the Institute for Scientific Information's Science Citation Index CD-ROM 2000, normalized by the size of a country's population. It measures the extent to which national researchers are working at the world-class level in S&T.

15. The amount and reliability of such statistics cannot (necessarily) be blamed on the national statistical institutes. It is inherent in the nature of developing economies that many economic activities occur outside the statistical purview (for example, payments in kind and production for personal, local, or regional consumption).

16. For a technical note on how the UNDP calculates the Human Development Index, see (http://hdr.undp.org/en/media/hdr_20072008_tech_note_1.pdf).

17. SPSS was used to conduct this analysis (SPSS Inc., Chicago, Ill.).

18. Contact the author for more information.

INDEX

Access to information: benefits of network membership, 37–38; economic development and, 25; feedback loops, 24; global collaboration in science policy-making, 110–12; intellectual property protection and, 73, 99–100; in nondirectional network, 37; open access, 6–7, 108; print resources, 71, 72–73; scientific capacity, 85–86, 87; scientific nationalism and, 73
Adaptive systems, 34–35
Akcelrud, D. E., 70–71
Albert, R., 42
Argonne National Laboratory, 44
Asia-Pacific Economic Cooperation forum, 91
Astronomy, 5–7, 76
Astrophysics, 56, 57
Authorship of scientific papers, 42, 53–55; indicators of scientific capacity, 87–88
Axelrod, R., 35

Barabási, A-L., 42
Bayh-Dole Act, 99
Beaver, D. deB., 21
BeppoSAX, 5–7
Bernal, J. D., 103
Bin Hu, 70
Biochemistry, 35
Boyle, Robert, 18–19, 46–47
Brazil, 15–18
Bush, Vannevar, 23–24, 89
Butterfield, H., 11, 20

Ca, Tran Ngoc, 89, 90, 91, 92, 93
Canada, 89
Capacity. *See* Scientific capacity
Cardoso, G., 15
Castells, M., 15
Center for Nanoscale Materials, 44
Centre National de la Recherche Scientifique, 3
CERN (European Organization for Nuclear Research), 26, 73